I0020764

BBC Micro:bit in Practice

A hands-on guide to building creative real-life projects with
MicroPython and the BBC Micro:bit

Ashwin Pajankar

Abhishek Sharma

Sandeep Saini

BIRMINGHAM—MUMBAI

BBC Micro:bit in Practice

Copyright © 2022 Packt Publishing

All rights reserved. No part of this book may be reproduced, stored in a retrieval system, or transmitted in any form or by any means, without the prior written permission of the publisher, except in the case of brief quotations embedded in critical articles or reviews.

Every effort has been made in the preparation of this book to ensure the accuracy of the information presented. However, the information contained in this book is sold without warranty, either express or implied. Neither the author(s), nor Packt Publishing or its dealers and distributors, will be held liable for any damages caused or alleged to have been caused directly or indirectly by this book.

Packt Publishing has endeavored to provide trademark information about all of the companies and products mentioned in this book by the appropriate use of capitals. However, Packt Publishing cannot guarantee the accuracy of this information.

Group Product Manager: Rahul Nair

Publishing Product Manager: Surbhi Suman

Senior Editor: Runcil Rebello

Technical Editor: Arjun Varma

Copy Editor: Safis Editing

Project Coordinator: Ashwin Kharwa

Proofreader: Safis Editing

Indexer: Pratik Shirodkar

Production Designer: Nilesh Mohite

Marketing Coordinator: Gaurav Christian

Senior Marketing Coordinator: Nimisha Dua

First published: December 2022

Production reference: 1171122

Published by Packt Publishing Ltd.

Livery Place

35 Livery Street

Birmingham

B3 2PB, UK.

ISBN 978-1-80461-012-1

www.packt.com

To Pandit Jawaharlal Nehru and Sardar Vallabhbhai Patel, the architects of modern India, whose names will always be remembered and cherished.

– Ashwin Pajankar

Contributors

About the author(s)

Ashwin Pajankar is an author, a YouTuber, and an instructor. He graduated from the International Institute of Information Technology, Hyderabad, with an MTech in computer science and engineering. He has been writing programs for over two and a half decades. He is proficient in Linux, Unix shell scripting, C, C++, Java, JavaScript, Python, PowerShell, Golang, HTML, and assembly language. He has worked on single-board computers such as Raspberry Pi and Banana Pro. He is also proficient with microcontroller boards such as Arduino and the BBC Micro:bit. He is currently self-employed and teaches on Udemy and YouTube. He also organizes programming boot camps for working professionals and software companies.

I want to thank my friend Anuradha who encouraged me to write this book. I thank the other two authors of the book, Abhishek and Sandeep. Finally, I would like to express my heartfelt gratitude toward the Packt team members – Neil, Runcil, Preet, Sayali, Surbhi, and Yogesh – for their valuable guidance and assistance.

Abhishek Sharma completed his BE in electronics engineering from Jiwaji University, Gwalior, India, and his PhD in engineering from the University of Genoa, Italy. He is presently working as an assistant professor in the Department of Electronics and Communication Engineering at the LNM Institute of Information and Technology, Jaipur, India. He is the coordinator of the ARM University Partner Program, Texas Instruments Lab, and Intel Intelligent Lab at the LNM Institute of Information and Technology and the center lead of **LNMIIT-Center of Smart Technology** (**L-CST**). His research interests are real-time systems and emerging technologies.

I would like to first and foremost thank my loving wife, Mano, and daughters, Madhu and Mihu, for their continued support, patience, and encouragement throughout the long process of writing this book. Thanks also to my parents for their support and faith.

Sandeep Saini works as an assistant professor in the Department of Electronics and Communication Engineering at the LNM Institute of Information and Technology, Jaipur, India. He has taught robotics and electronics subjects at the university level in India and abroad for over a decade. He has taught more than 8,000 students online and offline during this time. He received a BTech and MS in electronics and communication engineering from the International Institute of Information Technology, Hyderabad, India. He received his PhD from **Malaviya National Institute of Technology (MNIT)**, Jaipur, India. He has written 6 books and published 35 peer-reviewed journals and conference papers.

I would like to first and foremost thank my parents, my wife, and my daughter for their continued support, patience, and encouragement throughout the long process of writing this book.

About the reviewer

Emmanuel Efegodo is a software developer and an ed-tech practitioner with a degree in computer engineering. He enjoys programming generally but is mostly in love with JavaScript and the Jamstack architecture. He had his high school education in Gambia and his tertiary education in Nigeria. He strives to become an agent of change in Africa's digital education ecosystem, particularly in software design. He heads the curriculum team of JuniorX Innovation Academy with Obi Brown, a Google-certified ed-tech innovator, where he also teaches physical computing with the BBC Micro:bit, Python programming, and game development to young Africans.

Table of Contents

6

Interfacing External LEDs   95

7

Programming External Push Buttons, Buzzers, and Stepper Motors 125

Part 3: Filesystems and Programming Analog I/O

8

Exploring the Filesystem   145

Part 4: Advanced Hardware Interfacing and Applications

12

Producing Music and Speech   211

13

Networking and Radio   227

14

Advanced Features of the Micro:bit   241

15

Wearable Computing and More Programming Environments   253

Preface

The BBC Micro:bit is a very popular microcontroller board. It comes packed with features such as a display and various sensors. This board comes with a computing unit and the possibility to connect with various other peripheral devices. It is used by beginners to learn about the fundamentals of computer programming, electronics, and physical computing. It can be programmed using many programming frameworks such as MicroPython and Scratch.

The book covers many aspects of programming the BBC Micro:bit with the MicroPython programming language. The book begins with the basics and setup. Then, it covers the fundamentals of the Python programming language. After that, it explores various aspects of physical computing with the BBC Micro:bit, such as the programming of LEDs, buttons, buzzers, stepper motors, analog input, **Pulse Width Modulation** (**PWM**), internal sensors, and radio communication. It also explores topics such as the filesystem and wearable computing with a BBC Micro:bit.

After following the concepts, circuits, and code examples in this book, you will be comfortable with building your own project using the BBC Micro:bit with MicroPython as the preferred programming language.

Who this book is for

This book is for anyone who would like to use the combination of MicroPython and the BBC Micro:bit to build exciting real-life projects. Individuals working in domains such as embedded systems, electronics, software development, IoT, and robotics will find this book quite useful.

What this book covers

Chapter 1, *Introduction to the BBC Micro:bit*, covers the technical specifications of the BBC Micro:bit.

Chapter 2, *Setting Up the Micro:bit and Using Code Editors*, covers the installation of various integrated development environments for programming the Micro:bit with MicroPython.

Chapter 3, *Python Programming Essentials*, explores the basic concepts and syntax of Python programming.

Chapter 4, *Advanced Python*, dives deeper into the advanced concepts of Python.

Chapter 5, *Built-in LED Matrix Display and Push Buttons*, explores the programming of the built-in 5 x 5 matrix of LEDs and push buttons.

Chapter 6, Interfacing External LEDs, explores the interfacing and programming of external LEDs.

Chapter 7, Programming External Push Buttons, Buzzers, and Stepper Motors, teaches the programming of external output devices.

Chapter 8, Exploring the Filesystem, covers the built-in filesystem of the Micro:bit.

Chapter 9, Working with Analog Input and PWM, dives deeper into the interfacing of analog input devices. It also explores the PWM and interfacing of the output devices that employ PWM.

Chapter 10, Working with Acceleration and Direction, details working with internal sensors.

Chapter 11, Working with NeoPixels and a MAX7219 Display, teaches the interfacing of the external display.

Chapter 12, Producing Music and Speech, dives into producing music and speech with the Micro:bit.

Chapter 13, Networking and Radio, teaches you how to connect multiple Micro:bits together.

Chapter 14, Advanced Features of the Micro:bit, explores the advanced hardware features of the Micro:bit.

Chapter 15, Wearable Computing and More Programming Environments, covers simple projects in the area of sewable and wearable computing.

To get the most out of this book

You will need a BBC Micro:bit v2 for running the code examples. You will also need Thonny or Mu Editor installed on your computer—the latest versions, if possible. All code examples have been tested using Thonny and Mu Editor on Windows 10 and Linux. The code examples should work with future versions of the BBC Micro:bit, Thonny, Mu Editor, Windows, and Linux releases too.

Prior experience with some programming language, but not necessarily MicroPython, as well as building basic electronic circuits will be helpful when using this book.

Software/hardware covered in the book	Operating system requirements
BBC Micro:bit V2, Thonny, and Mu Editor	Windows, macOS, or Linux

If you are using the digital version of this book, we advise you to type the code yourself or access the code from the book's GitHub repository (a link is available in the next section). Doing so will help you avoid any potential errors related to the copying and pasting of code.

You should be comfortable with electronics and computer programming. Prior exposure to Python or MicroPython is desired but not mandatory.

Download the example code files

You can download the example code files for this book from GitHub at `https://github.com/ PacktPublishing/BBC-Micro-bit-in-Practice`. If there's an update to the code, it will be updated in the GitHub repository.

We also have other code bundles from our rich catalog of books and videos available at `https://github.com/PacktPublishing/`. Check them out!

Download the color images

We also provide a PDF file that has color images of the screenshots and diagrams used in this book. You can download it here: `https://packt.link/AYz3z`

Conventions used

There are a number of text conventions used throughout this book.

`Code in text`: Indicates code words in text, database table names, folder names, filenames, file extensions, pathnames, dummy URLs, user input, and Twitter handles. Here is an example: "This is because the `str1` and `pi` variables are not of the same data type."

A block of code is set as follows:

```
>>> print("This is a test string)
Traceback (most recent call last):
  File "<stdin>", line 1
SyntaxError: invalid syntax
```

Any command-line input or output is written as follows:

```
PS C:\Users\Ashwin> ufs ls
mylib.py main.py test.txt
```

Bold: Indicates a new term, an important word, or words that you see onscreen. For instance, words in menus or dialog boxes appear in **bold**. Here is an example: "Select **System info** from the **Administration** panel."

> **Tips or important notes**
> Appear like this.

Get in touch

Feedback from our readers is always welcome.

General feedback: If you have questions about any aspect of this book, email us at customercare@ packtpub.com and mention the book title in the subject of your message.

Errata: Although we have taken every care to ensure the accuracy of our content, mistakes do happen. If you have found a mistake in this book, we would be grateful if you would report this to us. Please visit www.packtpub.com/support/errata and fill in the form.

Piracy: If you come across any illegal copies of our works in any form on the internet, we would be grateful if you would provide us with the location address or website name. Please contact us at copyright@packt.com with a link to the material.

If you are interested in becoming an author: If there is a topic that you have expertise in and you are interested in either writing or contributing to a book, please visit authors.packtpub.com.

Share Your Thoughts

Once you've read *BBC Micro:bit in Practice*, we'd love to hear your thoughts! Scan the QR code below to go straight to the Amazon review page for this book and share your feedback.

https://packt.link/r/1804610127

Your review is important to us and the tech community and will help us make sure we're delivering excellent quality content.

Download a free PDF copy of this book

Thanks for purchasing this book!

Do you like to read on the go but are unable to carry your print books everywhere? Is your eBook purchase not compatible with the device of your choice?

Don't worry, now with every Packt book you get a DRM-free PDF version of that book at no cost.

Read anywhere, any place, on any device. Search, copy, and paste code from your favorite technical books directly into your application.

The perks don't stop there, you can get exclusive access to discounts, newsletters, and great free content in your inbox daily

Follow these simple steps to get the benefits:

1. Scan the QR code or visit the link below

https://packt.link/free-ebook/9781804610121

2. Submit your proof of purchase
3. That's it! We'll send your free PDF and other benefits to your email directly

Part 1:
Getting Started
with the BBC Micro:bit

This part aims to introduce you to the hardware and software. The first chapter introduces the specifications of the BBC Micro:bit, which is controlled by user-written code. The second chapter helps you set up the environment for the code. The third and the fourth chapters explore the Python programming syntax in detail. These chapters will help you follow and practice the book's concepts.

This part has the following chapters:

- *Chapter 1, Introduction to the BBC Micro:bit*
- *Chapter 2, Setting Up the Micro:bit and Using Code Editors*
- *Chapter 3, Python Programming Essentials*
- *Chapter 4, Advanced Python*

1

Introduction to the BBC Micro:bit

I certainly hope that you have read the preface and the table of contents, which provide a fair idea about our journey into the amazing world of the **BBC Micro:bit** (also written as **Micro:bit** or **Micro Bit**). This introductory chapter will warm you up for the upcoming exciting journey into the vast world of the Micro:bit. The road ahead is full of interesting concepts and projects. It is always a good idea to prepare well for the journey ahead, and this chapter will accomplish that.

We will explore the following topics in this chapter:

- The history of the Micro:bit
- The specifications of Micro:bit V1 and Micro:bit V2
- Powering up the Micro:bit
- Breakout boards
- Fritzing to create circuit diagrams

Let's get started!

Technical requirements

We will need the following hardware for this chapter:

- BBC Micro:bit V1 or V2
- A computer with Windows, macOS, or Linux
- A BBC Micro:bit edge connector
- A Micro-USB to USB cable

- An internet connection

- A mobile power bank

- Kitronik Mi:power

The history of the Micro:bit

It is important to know the history of the BBC Micro:bit. The **British Broadcasting Corporation** (**BBC**) is the United Kingdom's public broadcaster. It is also the world's oldest and biggest broadcaster. BBC has always been pioneering in creating programs for outreach in science and technology to improve the public understanding of science. Its programs include various documentaries and television series.

One such interesting television series was *The Computer Programme*. It was broadcast on BBC Two and used a home computer, the BBC Micro, conceptualized by the BBC and developed by **Acorn Computers**. The TV series was a part of the **BBC Computer Literacy Project**. The BBC Micro had six different models, which were all based on the famous MOS Technology 6502 8-bit microprocessor. It is a simplified and faster version of the Motorola 6800 microprocessor. The 6502 is a very popular microprocessor, and variants of it were used in popular video game consoles and computers such as Atari 2600, Apple II, Nintendo Entertainment System (popularly known as NES or Famicom), Commodore 64, and, of course, the BBC Micro. The BBC Micro was very successful, and it made a great impact in the computer education sector, leaving a great legacy behind.

In 2012, with the release of **Raspberry Pi**, a new era was ushered into the world of computing and education. Through *the Computer Literacy Project*, the BBC sought to build upon the legacy of the BBC Micro after the great success of Raspberry Pi. It onboarded many partners from industry, such as Microsoft, and academia, such as Lancaster University. The first version (now referred to as the **Micro:bit V1**) was launched in July 2015 and was available for general sale in March 2016. The BBC also gave hundreds of thousands of Micro:bits to school children in the UK as a part of science education outreach. After the Micro:bit successfully launched, the BBC formed a *not-for-profit* organization known as the **Microbit Foundation**.

> **Note**
> For more details, you can visit the home page of the Microbit Foundation at https://microbit.org/.

In October 2020, the Microbit Foundation released the second version of the Micro:bit. V2 has got better specifications than V1 at the same price. We will explore the specifications of V1 and V2 side by side in the following section.

The specifications of Micro:bit V1 and Micro:bit V2

The following table compares the features of the BBC Micro:bit V1 and V2 side by side (source: https://microbit.org/):

BBC Micro:bit	V1	V2
Processor	Nordic nRF51822	Nordic nRF52833
Flash memory	256 KB	512 KB
RAM	16 KB	128 KB
Speed	16 MHz	64 MHz
Bluetooth	Bluetooth 4.0	Bluetooth 5.1 with Bluetooth Low Energy (BLE)
Radio communication	2.4 GHz radio (80 channels)	2.4 GHz radio (80 channels)
Buttons	Two programmable (A and B) and one system (reset)	Two programmable (A and B) and one system (power/reset)
On/off switch	None	Press and hold the rear power button
Touchpad	None	Touch-sensitive logo
Microphone	None	Onboard Knowles SPU0410LR5H-QB-7 MEMS microphone (with LED indicator)
Display	5x5 programmable LED matrix (25 LEDs in total)	5x5 programmable LED matrix (25 LEDs in total)
Speaker	None	Onboard JIANGSU HUANENG MLT-8530 (up to 80 dB)
Motion sensor and compass	LSM303AGR	LSM303AGR
Temperature sensor	On-board temperature sensor	On-core NRF52
Edge connector	25 pins	25 pins

Table 1.1 – Comparison of the features of BBC Micro:bit V1 and V2

The processors used in both versions are a special type of processor known as a **System on Chip** (also abbreviated as **SoC** or **SOC**). An SoC is an **Integrated Circuit** (**IC**) that has all or most components of a complete working computer system. A typical SoC has a processor, flash memory, and RAM. Both versions employ **Advanced RISC Machines** (**ARM**) processors. The ARM uses **Reduced Instruction Set Computer** (**RISC**) instruction architecture. The V1 uses Nordic nRF51822 SoC (https://www.nordicsemi.com/Products/nRF51822), and the V2 uses Nordic nRF52833 (https://www.nordicsemi.com/products/nrf52833). The **Random Access Memory** (**RAM**) is used to execute the programs. The flash memory is used to store the programs, and it is reprogrammable.

The following diagram shows the front faces of V2 and V1 side by side:

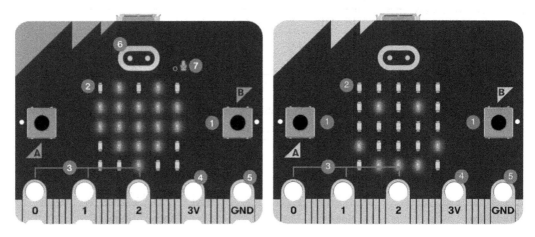

1. Buttons
2. LED display and light sensor
3. General-purpose input output pins
4. 3 volt power pin

5. Ground pin
6. Touch logo (V2 only)
7. LED for microphone (V2 only)

Figure 1.1 – Hardware features on the fronts of V2 and V1 (courtesy:
©Micro:bit Educational Foundation/microbit.org)

The following diagram shows the rear of the V2 and V1 side by side:

1. Radio and Bluetooth antenna
2. Processor and temperature sensor
3. Compass
4. Accelerometer
5. Pins
6. Micro USB socket
7. Single yellow LED
8. Reset button
9. Battery socket
10. USB interface chip
11. Speaker – V2
12. Microphone – V2
13. Red power LED – V2
14. Yellow USB LED – V2
15. Reset and power button – V2

Figure 1.2 – The hardware features on the rears of V2 and V1 (courtesy:
©Micro:bit Educational Foundation/microbit.org)

Another important aspect of the Micro:bit boards of both versions is that they come with **edge connectors** to interface with external hardware components. The following diagram explains the slight difference between the edge connectors of V2 and V1 side by side:

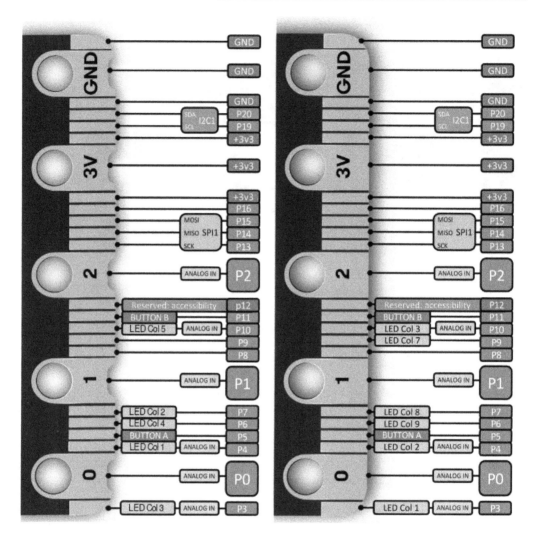

Figure 1.3 – The edge connectors of V2 and V1 (courtesy: https://
tech.microbit.org/hardware/edgeconnector/)

You can read online about the edge connectors in detail at `https://tech.microbit.org/hardware/edgeconnector/` and `https://microbit.pinout.xyz/`.

I understand that you may feel a bit overwhelmed with all this technical information at this stage. Without enough context about the utility of this technical information, it is natural to feel that way. However, in the upcoming chapters, we will learn about and demonstrate all these features in detail.

Now that we have a fair understanding of the history and specifications of the Micro:bit, let's learn various methods to power it up.

Powering up the Micro:bit

There are a few ways we can power up the Micro:bit. Let's see them all one by one. The following diagram clearly shows the micro-USB port and the battery socket. We can power up the BBC Micro:bit using these:

Figure 1.4 – The battery socket and micro-USB port (courtesy: https://commons.
wikimedia.org/wiki/File:BBC_micro_bit_%2826238853955%29.png)

We can use a micro-USB male to USB male cable to power the Micro:bit. The following is the micro-USB end of such a cable:

Figure 1.5 – A micro-USB male connector (courtesy: https://commons.
wikimedia.org/wiki/File:MicroB_USB_Plug.jpg)

Insert this end into the Micro:bit, as shown in the following photo:

Figure 1.6 – A micro-USB male connector (courtesy: https://commons.
wikimedia.org/wiki/File:Bbc-microbit-2021.jpg)

Insert the other end into a computer or a power bank. The following is an image of a mobile/portable power bank:

Figure 1.7 – A power bank with a micro USB cable attached (courtesy: https://
commons.wikimedia.org/wiki/File:Portable_power_bank.jpg)

We can also use a pair of AAA batteries with a special connector, as shown in the following photo:

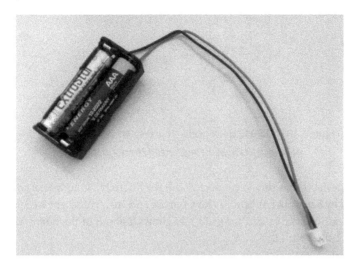

Figure 1.8 – A battery connector (courtesy: https://commons.wikimedia.org/wiki/File:Cavo_Microbit.jpg)

You can procure such a connector online at various marketplaces. One such page is https://www.sparkfun.com/products/15101. There are many other websites too that sell these connectors. You can also check your local makers' electronic supply shops for this.

The following photo shows the Micro:bit powered up with this connector and a pair of AAA batteries:

Figure 1.9 – A battery connector connected to the Micro:bit

We can also power the Micro:bit with a **CR2032**-type power cell, as shown in the following figure:

Figure 1.10 – CR2032 power cells (courtesy: https://commons.
wikimedia.org/wiki/File:Cr2032-7mmgrid.jpg)

We can use various special connectors to connect it with the Micro:bit. One special connector board is MI – the *power board* by **Kitronik** (https://kitronik.co.uk/products/5610-mipower-board-for-the-bbc-microbit). *Figure 1.11* shows a photo of the board, the nuts, and the CR2032 battery that comes with it:

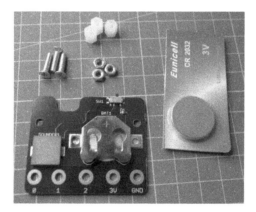

Figure 1.11 – The MI:power board and contents of the package

Figure 1.12 is a photo of the rear of the Micro:bit attached to the MI:power board:

Figure 1.12 – The Micro:bit assembled with the MI:power board

We can see that there is a dedicated **ON/OFF** switch. Attaching the board to the Micro:bit is very easy, and we can check the instructions for assembly at `https://kitronik.co.uk/products/5610-mipower-board-for-the-bbc-microbit`.

Both versions of the Micro:bit come with booklets, as shown in the following image:

Figure 1.13 – Micro:bit instruction booklets

It is recommended to go through them for a better understanding.

The out-of-box experience

When we unbox the Micro:bit and power it up for the very first time, it runs a factory default program known as the **out-of-box experience**. It is lots of fun to learn about the features of the Micro:bit using this program. Unbox and power your Micro:bit to run this program. Enjoy exploring the features of the Micro:bit.

Now that we have explored various ways to power the Micro:bit, we will explore special hardware components known as **breakout boards**.

Breakout boards

The **General Purpose Input Output (GPIO)** pins of the Micro:bit are extremely narrowly printed on the edge connector. It is difficult to use them directly, and soldering can ruin the board. So, many organizations have developed special products that make the GPIO pins of the Micro:bit easily accessible. These products are known by various names, such as breakout boards, GPIO expanders, I/O extensions, and edge connectors. The following are the URLs of the web pages for such products:

- `https://www.sparkfun.com/products/16445`

- `https://kitronik.co.uk/collections/microbit-and-accessories/`
  `products/5601b-edge-connector-breakout-board-for-bbc-microbit-`
  `pre-built`

- `https://robu.in/product/microbit-gpio-expansion-board/`

- `https://www.dfrobot.com/product-1867.html`

- `https://wiki.dfrobot.com/IO_Extender_for_microbit_V2.0_SKU_MBT0008`

I urge you all to procure one of these or any other edge connector of your choice, as we will need these for the demonstrations in this book.

I, too, own a couple of them, as shown in the following photos:

Figure 1.14 – A couple of edge connectors I own

Here, we have Micro:bit V1 and V2 inserted into the edge connectors:

Figure 1.15 – Micro:bits with the edge connectors

In the following section, we will get acquainted with a software program, **Fritzing**, that will help us visualize circuits.

Fritzing to create circuit diagrams

I am using software known as **Fritzing** (`https://fritzing.org/`) to create the circuits depicted in this book. It is not mandatory software for the demonstrations, as all the circuit diagrams are already printed in the book. However, if you wish to create your own circuit diagrams with Fritzing, you must procure them separately for 8 euros from `https://fritzing.org/download/`. I have included the Fritzing diagram files (with the `.fzz` extension) in the code bundle of the book. You can open them using the Fritzing software and modify them. It is very convenient software, and many hardware hackers (including us, the book's authors) use it to design and visualize their projects.

Fritzing has a library of many routinely used electronic and electrical components that include various boards. Also, users can create their own custom components such as boards and add them to their Fritzing setup. Many of them make such components available to other users for free. These components are stored in files with the `.fzpz` extension. You can find the component files for the BBC Micro:bit and many edge connectors at the following URLs:

- `https://tech.microbit.org/docs/hardware/assets/Microbit.fzpz.zip`
- `https://forum.fritzing.org/t/improved-micro-bit-part/7288`

Download the part files. Open the Fritzing software. In the **Parts** panel (the top-right panel), right-click on an empty gray space to show a dropdown, as shown in *Figure 1.16*:

Figure 1.16 – Importing a part to Fritzing

The first option is **Import…**. Click on that, and it opens a standard file selection window of the OS. Select the downloaded component files for the Micro:bit and edge connectors (those with the .fzpz extension) and import them. We can import only one component at a time. Once imported, all the components will be visible in the tab labeled **MINE** (refer to *Figure 1.16*). The following figure is a screenshot of those components added to a circuit diagram under development:

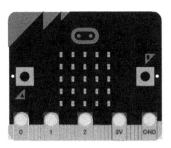

Figure 1.17 – Using the BBC micro:bit and edge connector parts in circuit diagrams

This is how Fritzing can be used to visualize the circuits we will build for the demonstrations throughout this book. You will find similar and more detailed circuit diagrams in the upcoming chapters of the book.

Summary

In this chapter, we learned a few fundamentals about the BBC Micro:bit. We had a brief tour of the hardware features that we will explore in the coming chapters. We also learned about the ways to power the board and edge connectors.

In the next chapter, we will focus on the software aspect of the Micro:bit and learn the basics of Python and MicroPython. We will start with installing various IDEs. We will also learn how to upgrade the firmware of the Micro:bit.

Further reading

For more information, visit the following links:

- `https://microbit.org/`

- `https://www.elektormagazine.com/news/new-bbc-microbit-v2-speaker-microphone-touch-sensor`

2

Setting Up the Micro:bit and Using Code Editors

In the previous chapter, we got acquainted with the BBC Micro:bit board and its specifications. We also learned how to power it up. The chapter was heavy on the side of concepts and not the hands-on part.

In this chapter, we will get started with the environment setup for MicroPython with the Micro:bit. We will write and execute our first MicroPython program, and we will explore the entire ecosystem of MicroPython programming platforms and Integrated Development Environments (IDEs) with Micro:bit. We will also learn how to upgrade the firmware of Micro:bit. The following is the list of topics we will explore and demonstrate:

- BBC Micro:bit versus Raspberry Pi
- The Python programming language
- Introduction to MicroPython
- MicroPython code editors
- Using offline IDEs for MicroPython
- Manually upgrading the firmware
- Restoring the out-of-the-box experience program

Upon completing this chapter, we will be very comfortable with setting up BBC Micro:bit for MicroPython and running the basic code examples written in MicroPython. Let's get started.

Technical requirements

For this chapter, we will need a computer with an internet connection and a Micro:bit with a micro USB cable for connectivity.

BBC Micro:bit versus Raspberry Pi

Most of us have at least heard of, if not worked with, Raspberry Pi. Raspberry Pi is a family of single-board computers and microcontrollers developed by the Raspberry Pi Foundation (https://www.raspberrypi.com/). It was founded and headed by Eben Upton. The most recent single-board computer in the Raspberry Pi family, at the time of writing of this book (July 2022), is **Raspberry PI 4 Model B** and the latest microcontroller board is **Raspberry Pi Pico W**.

Raspberry PI 4 Model B is a full-fledged computer capable of running **operating systems** (**OSes**) such as **Windows IoT**, various **Linux** flavors, and **FreeBSD**. Raspberry Pi Pico W is a low-cost microcontroller board that can be programmed with similar tools that we use to program BBC Micro:bit. Since BBC Micro:bit is not a full-fledged computer like Raspberry Pi 4B, it makes very little sense to compare it with Raspberry Pi 4B and other single-board computers. However, the comparison between BBC Micro:bit V2 and Raspberry Pi Pico W seems to be a fair one. The following table compares a few of the features of both of them side by side:

	BBC Micro:bit V2	**Raspberry Pi Pico W**
SoC	Nordic nRF52833 – 64 MHz 32-bit ARM Cortex-M4 microcontroller	RP2040 Arm Cortex M0+ Dual Core at 133 MHz
Flash Memory	512 KB	2 MB
Static RAM	128 KB	264 KB
GPIO	26	40
Connectivity	Bluetooth and radio	Infineon CYW43439 2.4 GHz Wi-Fi with an onboard antenna connected via SPI
Power	micro USB, GPIO pins, and JST battery connector	micro USB and GPIO pins

Table 2.1 – Comparison between BBC Micro:bit V2 and Raspberry Pi Pico W

The main reason we see the comparison between BBC Micro:bit and Raspberry Pi Pico W is that both boards are capable of running MicroPython.

Similarly, you can compare various models of the Arduino family of microcontrollers with Micro:bit. Take it as an exercise. Start by comparing Micro:bit with the Arduino Uno R3, then explore the specifications and programming frameworks of other models of the members of the Arduino family.

In the next section, we will explore the concepts related to the Python programming language.

The Python programming language

It is important to know the background of the programming language we plan to use for our code demonstrations. Python (`https://www.python.org/`) is a general-purpose programming language, and it is free and open source. Python is also an object-oriented, high-level, and interpreted programming language. *Figure 2.1* is the logo of the Python (`https://www.python.org/community/logos/`) programming language:

Figure 2.1 – Python logo (courtesy: the Python Software Foundation)

Python is a dynamically-typed and garbage-collected programming language. Apart from object-oriented programming, it supports other programming paradigms such as functional, structured, and procedural programming. Python also comes with batteries included; this means it comes with a very large standard library.

The **Centrum Wiskunde & Informatica** (abbreviated as **CWI**, and in English, known as the **National Research Institute for Mathematics and Computer Science**) in the Netherlands was the birthplace of Python. **Guido van Rossum** is the creator of the Python programming language. It started as his personal hobby project in December 1989. He was the benevolent dictator for life for Python until 2018. He has worked for Google and Dropbox, and currently, he is a Distinguished Engineer at Microsoft. The following is a photograph of Guido van Rossum:

Figure 2.2 – Guido van Rossum (courtesy: https://commons.wikimedia.org/wiki/
File:Guido_van_Rossum_%286984267183%29_%28cropped%29.jpg)

Python is a successor of the ABC programming language, which was also conceived at the CWI. A steering committee currently guides Python development (as of now, Guido is not a part of it anymore).

The Python community follows **Python Enhancement Proposals** (**PEPs**) for the development and governance of the standards. We can find the index of PEPs at `https://peps.python.org`. The term PEP itself is defined in detail at `https://peps.python.org/pep-0001/`. Moreover, the philosophy of the Python programming language can be visited at PEP20 (`https://peps.python.org/pep-0020`).

Python implementations and distributions

We will extensively use Python syntax throughout the book to demonstrate various functionalities of Micro:bit. So, it is imperative to know what a Python implementation is. Before we begin to understand the meaning of the term *implementation*, let's talk a bit about UNIX and C. UNIX is a family of OSes; it is more like a standard, and there are many OSes, such as macOS, that are UNIX-certified. Also, there are many other OSes, such as FreeBSD and various Linux distributions, that emulate some of the functionality of UNIX OS. Similarly, C is a standard for programming languages. There are many compilers, such as GCC, LLVM, and VC++, that support compiling a C program. All we wish to say is that UNIX and C are more like standards, and they are implemented in many systems.

Similarly, Python is now becoming more of a standard. Python programs are interpreted by the Python interpreter. The reference implementation of this interpreter is available at www.python.org and it is known as **CPython**. Many other projects have created their own Python interpreters and they are known as Python implementations. We can find (non-exhaustive) lists of such implementations at `https://www.python.org/download/alternatives/` and `https://wiki.python.org/moin/PythonImplementations`.

Now, let's understand what a Python *distribution* is. An implementation of Python bundled with a package manager and a few libraries is known as a **Python distribution**. For example, a standard distribution of Python (available for download at www.python.org) has the reference interpreter **CPython**, the package manager (**pip** – **pip installs Python**, a recursive acronym – a technical pun), and an IDE (in this case, **Integrated Learning and Development Environment** – abbreviated as **IDLE**). There are many other distributions of Python developed by other organizations. A non-exhaustive list can be found at `https://wiki.python.org/moin/PythonDistributions`.

For this book and the demonstrations, we are more interested in the **MicroPython** implementation of Python.

Introduction to MicroPython

As its name indicates, MicroPython (`https://micropython.org/`) is an implementation of Python meant for microcontrollers. It is written in C and is largely compatible with Python 3. MicroPython is made up of a Python compiler that converts a program into bytecode and a runtime interpreter that runs the bytecode on the targeted microcontroller board. It also comes with a **Read-Eval-Print-Loop** (**REPL**) prompt for the immediate execution of Python statements. MicroPython comes with a large implementation of the core libraries included in CPython. *Figure 2.3* is MicroPython's latest logo:

Figure 2.3 – MicroPython's logo

MicroPython was created by the Australian programmer and theoretical physicist Damien George. It was financially supported by a Kickstarter crowdfunding campaign. It was originally meant for a specialized microcontroller board, **Pyboard**. Its scope has now expanded, and it supports tons of microcontroller boards, including both revisions of BBC Micro:bit. MicroPython has a small footprint for devices with 256 k of flash memory and 16 k of RAM. Micro:bit comes with built-in support for MicroPython, and we can directly run MicroPython programs on Micro:bit.

MicroPython code editors

There are several online and offline code editors for writing and uploading MicroPython code for microcontrollers. Let's explore a few of the most popular editors.

Online code editors

One of the best ways to start coding without installing anything is to use the online code editor at `https://python.microbit.org/`. Currently, there are two versions of this available, and we will explore both versions. The URL points to `https://python.microbit.org/v/3`. This is the current version of the online code editor, and in the future, it will be accessible at `https://python.microbit.org/v/3`. Let's explore it first. When we visit `https://python.microbit.org/`, the very first thing we see is the following interface:

Figure 2.4 – The online code editor interface

The web page is divided into many sections. We can see that an example code snippet is already present in the middle section. Do not worry about the code as we will learn about it all in detail throughout the book. The objective of this section is to familiarize ourselves with the online code editor. In case you are not comfortable with the font size, you can change it by clicking on the settings option (the gear symbol) in the bottom-left corner.

The code is error-free and works perfectly (we have already tried and tested it). In the top-right corner, there is a simulator of Micro:bit (V2, to be very specific), which allows us to test programs. If we click the **run** button (triangle enclosed in a circle) as shown in the following screenshot (*Figure 2.5*), then it runs the program in the simulator and shows the output:

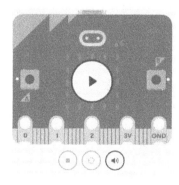

Figure 2.5 – The Micro:bit simulator

We can see that the two buttons at the bottom are disabled. They will be enabled once we click the **run** button. After we click the **run** button, it shows the execution of the code, as shown in *Figure 2.6*:

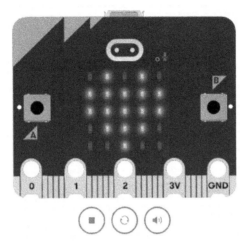

Figure 2.6 – The Micro:bit simulator running a program

We can see the output of the code on the built-in display of the simulator. We can also see that the buttons that were disabled earlier are enabled now. The first button is to stop the execution of the program, the second button is to reload the program, and the third button is to enable/disable the sound output.

At the bottom, there is a purple-colored button that reads **Send to Micro:bit**. When we click this button, it opens the following interface (*Figure 2.7*) to connect to a Micro:bit:

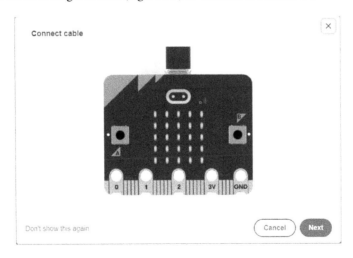

Figure 2.7 – Tutorial to connect Micro:bit to a computer

Connect your Micro:bit device to the computer using the cable as instructed on the window. Once you connect, the OS detects a new drive and opens the Micro:bit as a storage device, as shown in *Figure 2.8*:

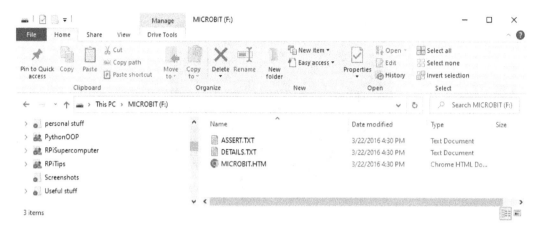

Figure 2.8 – Micro:bit as a storage device

One of the files on the Micro:bit drive is DETAILS.TXT. This file has valuable information about the board, such as the unique ID of the board and the firmware running on the board. The firmware enables the USB connection and helps the computer show Micro:bit as a storage drive. The contents of our Micro:bit V2 are as follows:

```
# DAPLink Firmware - see https://mbed.com/daplink
Unique ID: 9904360259304e45004890130000006a000000009796990b
HIC ID: 9796990b
Auto Reset: 1
Automation allowed: 0
Overflow detection: 0
Incompatible image detection: 1
Page erasing: 0
Daplink Mode: Interface
Interface Version: 0255
Bootloader Version: 0255
Git SHA: 1436bdcc67029fdfc0ff03b73e12045bb6a9f272
Local Mods: 0
USB Interfaces: MSD, CDC, HID, WebUSB
Bootloader CRC: 0x828c6069
Interface CRC: 0x5b5cc0f5
```

```
Remount count: 0
URL: https://microbit.org/device/?id=9904&v=0255
```

The contents will obviously be different for each board. For a detailed explanation of the contents, you may wish to visit https://support.microbit.org/support/solutions/articles/19000129907-details-txt.

The other file is MICROBIT.HTM, and it points to the page at https://microbit.org/get-started/user-guide/overview/.

As of now, we do not need to work with this window, so you can close it. If you look at the list of the drives in your OS, then you will find a drive labeled MICROBIT. In the window shown in *Figure 2.9*, click on the **Next** button and it will open the next page of the tutorial:

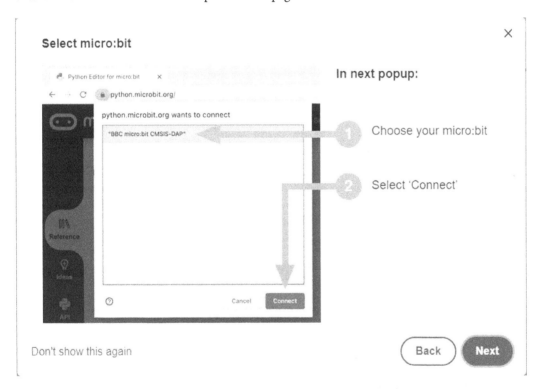

Figure 2.9 – Second screen of the tutorial window

When we click on the **Next** button, it opens a pop-up box, as shown in *Figure 2.10*:

Figure 2.10 – Connection pop up window

The pop-up box in the preceding figure shows all of the BBC Micro:bit devices connected to the computer. Select the device and click on the **Connect** button. Once it connects, the bottom menu will show additional options. If you click on the **Show Serial** option and then press *CTRL + C*, it shows the **REPL** window, as shown in *Figure 2.11*:

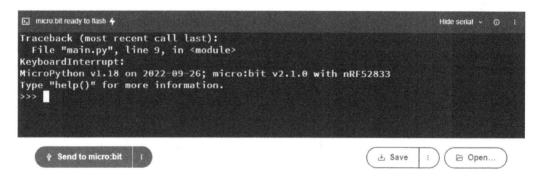

Figure 2.11 – The REPL window

We can use this REPL window to run small snippets of MicroPython code. When we click on the **Send to Micro:bit** button now, since it is connected to a Micro:bit device, it flashes the program to the Micro:bit device and the program starts executing on the device. We can see the text and the heart image scrolling and flashing on the 5 x 5 LED display matrix of the Micro:bit.

We can save the file in HEX format and as a Python program to the local disk by clicking on the **Save** button and three vertical dots beside that, respectively. We can also open a HEX or a Python code file using the **Open** button. Flashing the code to the Micro:bit device's flash memory using the **Send to Micro:bit** button is known as **Direct Flashing**.

> **Note**
> If you wish to know about the HEX file format, please explore the following URLs: `https://tech.microbit.org/software/hex-format/` and `https://microbit-micropython.readthedocs.io/en/v2-docs/devguide/hexformat.html`.

There is one other way of flashing the code to the Micro:bit. As explained earlier, we can save the code in HEX format and as a Python script to our local computer's disk. We can directly copy the HEX file (but not the Python program) to the drive labeled *MICROBIT*. First, download your program as a HEX file to your local computer. Now, copy the HEX file we downloaded to the *MICROBIT* drive. It will take some time for the file to copy (not too long, though, just a few seconds), and then the *MICROBIT* drive window will disappear immediately. This is because the drive is automatically disconnected and unmounted. However, the drive will automatically be connected and mounted in a couple of seconds, and the drive window will also appear automatically. If you check the drive's contents now, you won't find the HEX file. Do not panic! This is expected behavior. The program is loaded onto the Micro:bit's flash memory. We can see the text and the heart image scrolling and flashing on the 5x5 LED display matrix of the Micro:bit.

Using REPL

We can use **REPL** in MicroPython in the current live editor at `https://python.microbit.org/`. The future versions will also likely have a REPL prompt for MicroPython. REPL is similar to the command-line interface of any OS. We can use it to run single-line statements and small code blocks written in MicroPython. Type in `print("Hello, World")` and hit the **Enter** key on the keyboard to execute it. It will show the following result:

```
>>> print("Hello, World!")
Hello, World!
>>>
```

This way, we can run single-line statements and see the results immediately. Let's try another example. Run the following statements:

```
>>> from microbit import *
>>> display.scroll("BBC micro:bit V2")
```

Once both statements are executed, we can see the quoted text scrolling on the display of the Micro:bit. Don't worry much about the syntax right now. We will learn about everything in detail in the upcoming chapters. For now, you may wish to scroll the text of your choice by making changes to the quoted text and running it in the REPL prompt.

> **Note**
> The older version of this editor can be found at https://python.microbit.org/v/2.

In this section, we saw how we can use the online Python editor of Micro:bit to write and directly flash code to our Micro:bit devices.

Using offline IDEs for MicroPython

Python programs can also be written and executed with the help of specialized software known as IDEs. The reference implementation of Python (CPython) comes with an IDE known as IDLE. The following is a screenshot of IDLE in action (*Figure 2.12*) on a Windows computer:

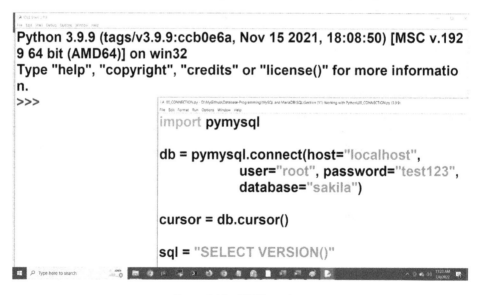

Figure 2.12 – IDLE in action

There is no doubt that IDLE has a great interface. However, it works with CPython and not with MicroPython. We have already seen how we can work with web editors for MicroPython. In this section, we will explore a few desktop IDEs for MicroPython.

The Thonny Python IDE

Let's go over how to install and use the Thonny Python IDE. We can visit `https://thonny.org/` for the installation files. In the top-right corner of the page, we have options for Windows, macOS, and commands for installation on distributions of Linux. Thonny comes bundled with a Python interpreter of its own. It may not conform to the latest CPython release from Python Foundation and could be older. As of the writing of this book, the latest CPython release is 3.10.5, and the latest stable version of Thonny comes with Python 3.7.9.

Thonny comes pre-installed with the Raspberry Pi OS if you choose to use a desktop environment. You can install it on any distribution of Linux with the following command:

```
bash <(wget -O - https://thonny.org/installer-for-linux)
```

If you have already installed an implementation of Python 3 with pip3 and wish to only install Thonny, then run the following command at the command prompt of your OS:

```
pip3 install thonny
```

We prefer this method. We downloaded and installed the latest Python 3 setup from `www.python.org` and then ran the preceding command on Windows PowerShell.

For Debian and its derivatives, the following command will install Thonny:

```
sudo apt install thonny
```

For Fedora and its derivatives, the following command will install Thonny:

```
sudo dnf install thonny
```

Let's see how we can work with Thonny. You can launch it from the desktop shortcut or the command prompt of your OS using the following command:

```
thonny
```

Once launched, it shows the following window:

Figure 2.13 – The Thonny IDE window

We can see the menu, code editor, assistant, and shell. The shell window provides REPL functionality. In the bottom-right corner, you can find the Python interpreter associated with the current instance of Thonny. It is a clickable option and will invoke the following options when clicked:

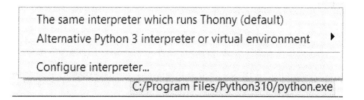

Figure 2.14 – Interpreter options

Click on the last option (**Configure interpreter…**) and it will open the following window:

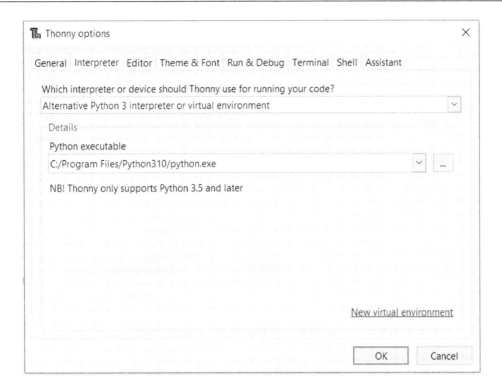

Figure 2.15 – Configuring the interpreter...

We can also open this window from the main menu from the **Tools** option, as shown in *Figure 2.16*:

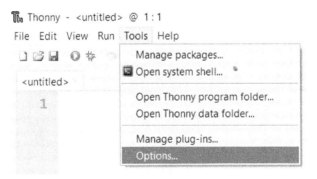

Figure 2.16 – Tools | Options

The second tab of the window will open the interpreter options (refer to *Figure 2.15*). Here, we can choose to associate the Thonny IDE with a particular interpreter. If we choose **Alternative Python 3 interpreter or virtual environment** from the first dropdown, as shown in *Figure 2.15*, then we can mention the path of the Python 3 interpreter executable in the text area (we can also browse and choose it). This way, we can use any implementation of Python with Thonny. We are more interested in using Thonny with MicroPython using BBC Micro:bit. First, connect your Micro:bit to your computer and then choose the **MicroPython (BBC Micro:bit)** option from the first dropdown (*Figure 2.17*):

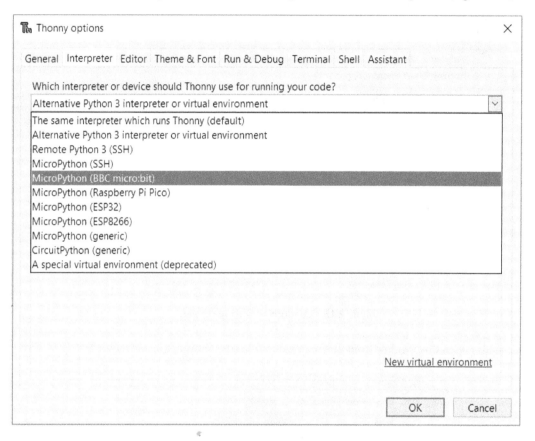

Figure 2.17 – Configuring for MicroPython with Micro:bit

In the second dropdown, we have to choose the port. Look for an option that has **USB Serial** or **UART** in the text and choose that option:

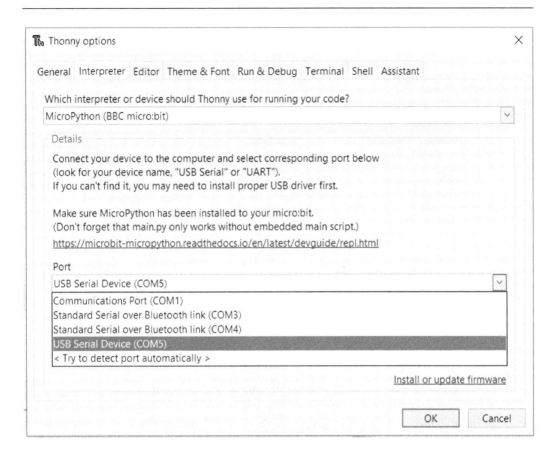

Figure 2.18 – Choosing the correct port

We can install the firmware using this tool. Click on **Install or update firmware** in the bottom-right corner of the window. It opens a new window, as shown in *Figure 2.19*:

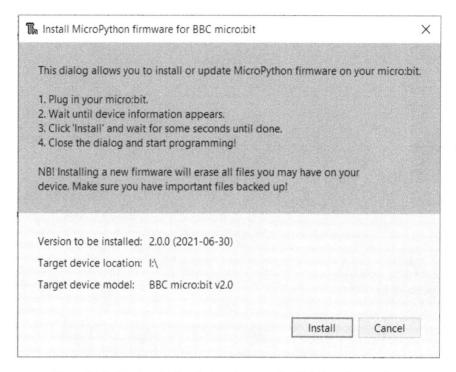

Figure 2.19 – The Install MicroPython firmware for BBC Micro:bit window

Click on the **Install** button and it will install/upgrade the device's firmware. Click on the **OK** button and the shell will restart and show the following header indicating the detection of Micro:bit board:

```
MicroPython v1.15-64-g1e2f0d280 on 2021-06-30; micro:bit v2.0.0 with nRF52833
Type "help()" for more information.
>>>
```

Figure 2.20 – Thonny configured for Micro:bit

You may wish to try the following two lines again:

```
>>> from microbit import *
>>> display.scroll("BBC micro:bit V2")
```

Also, copy and paste the same code in the editor window and save that as prog00.py to a location on your computer. We prefer saving the code in the directories organized chapter-wise. You can download all these examples from the GitHub repository for this book project. When you try to save any new code file for the very first time, it shows the following window:

Figure 2.21 – Options to save

Choose the **This computer** option and save it to a local location. This is the copy on the local computer. Now, save again and this time, choose the second option (**Micro:bit**). This will open a new window, as shown in *Figure 2.22*:

Figure 2.22 – Saving on BBC Micro:bit

Enter `main.py` in the text area corresponding to the **File name** label and click on the **OK** button. The code has been flashed to the connected Micro:bit. However, it won't run immediately. We have to reboot the Micro:bit. You can either disconnect and reconnect it again (hard reboot) or click the REPL shell and press *CTRL + D* on the keyboard simultaneously (soft reboot). We prefer the latter.

If we want a MicroPython program to run when Micro:bit is rebooted, we should save the file as `main.py`. If we modify the program, we should save the local copy and overwrite the `main.py` file on the BBC Micro:bit.

The Mu editor

Let's see how to install and configure the **Mu editor**. Please visit the page at `https://codewith.mu/en/download` and download it for your platform. You can also install it with pip3 using the following command:

```
pip3 install mu-editor
```

Once installed, launch the Mu editor using the shortcut or typing in the following command in the command prompt:

```
mu
```

It will launch the following interface:

Figure 2.23 – The Mu editor

The Mu editor is currently configured for Python 3. We can check it in the bottom right of the preceding screen. If we click the very first icon (the one that reads **Mode**), it opens the following dialog box (*Figure 2.24*) where we can choose the interpreter:

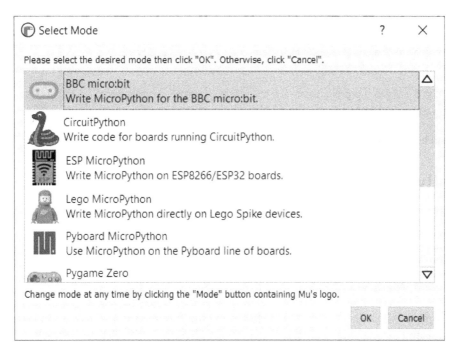

Figure 2.24 – Choosing the interpreter

Choose the option that reads **BBC Micro:bit**, and click the **OK** button. We are ready to roll. We can see the interpreter's name on the bottom right. Also, we can see that a couple of options in the main menu have changed. Load the code file that we saved earlier to the local disk of our computer and click the **Flash** option. It will directly flash the code to the Micro:bit and run it (unlike Thonny, there is no need to save it manually in the Micro:bit and soft/hard reboot). There is also an option for REPL. As an exercise, explore that. This is how we can use the Mu editor to code with MicroPython for Micro:bit.

Working with other editors and smartphone apps

Another popular IDE is **uPyCraft**. You can explore it at its GitHub page at `https://github.com/DFRobot/uPyCraft`.

We can check the information about other editors and apps at `https://support.microbit.org/support/solutions/19000102309`.

The following page has information about the Android app:

`https://support.microbit.org/support/solutions/articles/19000065804-using-the-micro-bit-on-android`

The following pages have information about the iOS app:

- `https://support.microbit.org/support/solutions/articles/19000117215-micro-bit-ios-app-connect-and-send-programs`

- `https://support.microbit.org/support/solutions/articles/19000117216-monitor-control-in-the-ios-app`

Next, we will learn how to manually upgrade the firmware of a Micro:bit board to keep it up to date.

Manually upgrading the firmware

A firmware is a program between the hardware and the software; it provides the control of the device-specific hardware. A well-known example is **Binary Input Output System (BIOS)**.

In the Micro:bit drive, the `DETAILS.TXT` file has the firmware number in the following line:

```
Interface Version: 0255
```

Earlier, we updated the Micro:bit firmware before using the Thonny editor. Now, we will learn how to do it manually. First, we have to identify the version of Micro:bit. A detailed article teaches us how to identify the version of Micro:bit and you can find it at `https://support.microbit.org/support/solutions/articles/19000119162-how-to-identify-the-version-number-of-your-micro-bit-`. The following diagram (*Figure 2.25*) shows the difference between V1 and V2 markings:

Figure 2.25 – V1 versus V2 markings (courtesy: Micro:bit Educational Foundation/microbit.org)

All of the V1 boards use the same firmware.

There are two versions for V2. One version is for the boards numbered V2.0, and another version is for the boards numbered V2.20 or V2.21. The following figure shows the difference between both versions of V2:

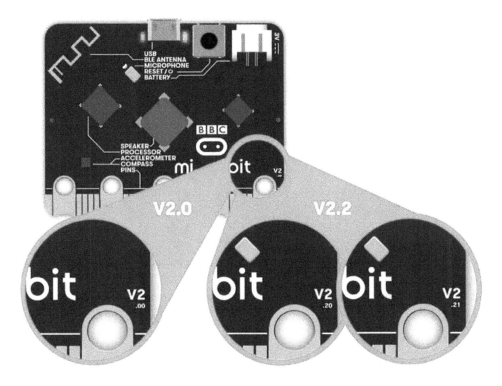

Figure 2.26 – V2.0 versus V2.2x (courtesy: Micro:bit Educational Foundation/microbit.org)

We can download the HEX file for the V1 firmware from https://tech.microbit.org/docs/software/assets/DAPLink-factory-release/0249_kl26z_microbit_0x8000.hex.

We can download the HEX file for the V2.0 firmware from https://tech.microbit.org/docs/software/assets/DAPLink-factory-release/0255_kl27z_microbit_0x8000.hex.

We can download the HEX file for the V2.20 and V2.21 firmware from https://tech.microbit.org/docs/software/assets/DAPLink-factory-release/0257_nrf52820_microbit_if_crc_c782a5ba90_gcc.hex.

We can find all these URLs at https://microbit.org/get-started/user-guide/firmware/.

Download the HEX file for your version of Micro:bit. Then, keep the push button near the micro USB port pressed and connect the Micro:bit to the computer:

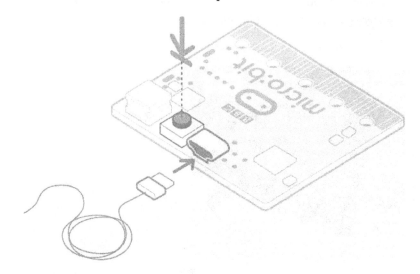

Figure 2.27 – Connecting to upgrade the firmware (courtesy:
Micro:bit Educational Foundation/microbit.org)

Rather than mounting normally as a **MICROBIT** drive, it will mount as a **MAINTENANCE** drive (*Figure 2.28*) as follows:

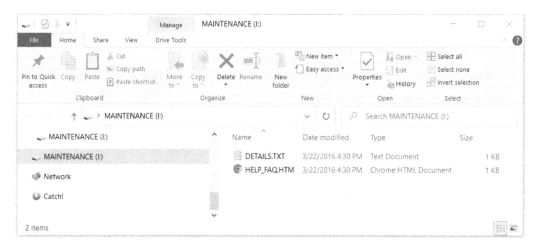

Figure 2.28 – Maintenance drive

Copy the downloaded HEX file to this **MAINTENANCE** drive and it will upgrade the firmware. The Micro:bit will disconnect and reconnect automatically.

Restoring the out-of-the-box experience program

We have overwritten the factory out-of-the-box experience program by writing our own programs to the Micro:bit. If we wish to reload the out-of-the-box experience program, we can find its HEX file at `https://microbit.org/get-started/user-guide/out-of-box-experience/`.

Summary

In this chapter, we compared Micro:bit with Raspberry Pi Pico W. We learned the history of the Python programming language and MicroPython. We explored various IDEs for MicroPython and ran our first MicroPython program with Micro:bit. We also learned how to upgrade the firmware and restore the out-of-the-box experience. MicroPython is one of the best programming platforms for beginners to work with Micro:bit and similar microcontroller boards, such as Raspberry Pi Pico W. All the demonstrations in the coming chapters of this book will use MicroPython.

In the next chapter, we will explore the basics of Python programming with MicroPython and Micro:bit in depth. The chapter will be heavy with hands-on Python programming. Readers who have never programmed with Python before will learn a lot, and experienced Python readers will also be able to refresh their understanding of the Python programming basics.

Further readings

We can explore the information about the IDEs used for MicroPython at their project home pages:

- `https://codewith.mu/`
- `https://thonny.org/`
- `https://www.jetbrains.com/pycharm/`
- `https://dfrobot.gitbooks.io/upycraft/content/`
- `https://python.microbit.org`

We also recommend that all readers read the Python language official documentation at `https://docs.python.org/3/`.

3

Python Programming Essentials

In the previous chapter, we got acquainted with Python's history and implementations. We also learned how to use an online editor by the Micro:bit Foundation for programming a Micro:bit device (also referred to as a board). We installed the Thonny and Mu IDEs for MicroPython programming and explored their useful features for programming Micro:bit with MicroPython. I hope that you have had an opportunity to write a basic program for Micro:bit.

In this chapter, we will get started with hands-on programming with MicroPython. We will get comfortable with the Python programming syntax before getting into creating projects with MicroPython and Micro:bit. If you are new to Python programming, this chapter will help you to gain the essentials. We will explore the following list of topics together:

- Getting started with Python programming
- Conditional statements
- Loops
- Computing prime numbers, factorials, and Fibonacci series

Let's get started with the basics of programming.

Technical requirements

We do not need anything else besides a computer, a Micro:bit device, and a micro USB cable for the demonstrations in this chapter.

Getting started with Python programming

A program is a set of instructions. A computer runs those instructions in the given sequence and produces an output (also known as a result). In this section, we will take the first steps toward writing our own small programs with the Python syntax. If you recollect, I have mentioned in the previous chapter that the MicroPython implementation is largely compatible with Python 3. In this chapter, we will explore the common syntax for Python 3 and MicroPython. To keep it simple, we won't be exploring any features of Micro:bit and specialized MicroPython syntax.

In the previous chapter, we ran simple statements on REPL with the online Python editor, Thonny, and Mu. I prefer using the Thonny editor and will continue using it for the rest of the book, but you can use the IDE of your choice. Open any of these editors, and then open the REPL interface. I hope that you are already comfortable working with this simple interface. Let's use it to further enhance our knowledge of Python programming. We can write simple instructions in REPL (also known as the Python shell) to execute them immediately. In the terminology of programming languages, these instructions are known as statements. The following code snippet is a simple statement executed in the Python shell:

```
>>> 2 + 2
4
```

Try it now. Try other mathematical operators such as -, *, /, and %. The following are examples of these:

```
>>> 2 - 2
0
>>> 2 / 2
1.0
>>> 2 % 2
0
>>>
```

The numbers we are using are operands. I have only used one number with the operands, however, we can (and should) try other numbers too. And why limit ourselves to integers when we can also use numbers with a decimal point? Such numbers are known as **floating-point numbers** or **floats**. The following are examples of floating-point arithmetic:

```
>>> 22/7
3.142857
>>> 1.71828 + 1
2.71828
>>> 10.807 - 1
9.807
```

Well, that brought back memories of my school science lessons. We have already written the customary `Hello, World` program in the previous chapter. So, let's try something else now:

```
>>> print(2)
2
```

```
>>> print(2+2)
4
```

Let's print a string:

```
>>> print("This is a test string)
Traceback (most recent call last):
  File "<stdin>", line 1
SyntaxError: invalid syntax
```

Aghrrrr! I did it deliberately. The invalid syntax error is generated because we missed closing the string with a double quote. Python interpreters classify this as a syntax error. Let's rectify and run it again:

```
>>> print("This is a test string")
This is a test string
```

We can also use single quotes to specify a string as follows:

```
>>> print('This is a test string')
This is a test string
```

Let's understand the concept of **Boolean values**. We can have a special variable that stores the truth value in programming terminology. These values are known as Boolean values (a tribute to *George Boole*), and such a variable is a Boolean variable (or Boolean data type). There are only two Boolean values, True and False:

```
>>> True
True
>>> False
False
```

We can perform logical operations as follows:

```
>>> True and True
True
>>> True and False
False
>>> True or False
True
>>> True or True
```

```
True
>>> not True
False
>>> not False
True
>>>
```

These are logical operations. `True` and `False` are keywords and are case sensitive. If you make a mistake, the interpreter returns the syntax error as follows:

```
>>> not true
Traceback (most recent call last):
  File "<stdin>", line 1, in <module>
NameError: name 'true' isn't defined
```

In the next section, we will understand and experiment with the concept of variables in detail.

Variables

We have directly handled strings, numbers, and Boolean values. We can store them in the main memory of the computer (RAM), where programs are executed. We can also assign them names to refer to them again when needed. These named storage areas in RAM are known as **variables**. We can only refer to them from the program in which they are created. Once the program stops running, they cannot be accessed. Also, they are limited to the program that created them even when the program is running, and outside programs cannot access them directly.

If you have worked with C, C++, Java, and other programming languages, you will know that you have to declare the data type of the variable in advance, as follows (the sample C code snippet won't be executed by the Python interpreter):

```
int a = 2;
float pi = 3.14;
```

In these languages, after creating a variable with a particular data type, we cannot change its data type. Python is very flexible in that regard. We do not have to declare data types, and we can change the data type of any variable on the fly by assigning it a value of that particular data type as follows:

```
>>> a = 1
>>> a
1
>>> a = 3.14
```

```
>>> a
3.14
>>> a = 'c'
>>> a
'c'
>>> a = 'Hello, World!'
>>> a
'Hello, World!'
```

Try that one. And also, try changing the value of a again.

Now, let's run a few statements:

```
>>> a = 1
>>> b = 2.14
>>> print(a + b)
3.14
>>> print("My first script!!")
My first script!!
```

Now, let's collect all four statements and copy and paste them into the editor. Save the file in a directory and name the file prog00.py. Thonny automatically assigns it a py extension. I prefer to maintain chapter-wise directories. I recommend you do the same. Refer to the code bundle of this book for more information. After saving the program, click on the **Run** (green triangle) button in the menu, as shown in the following screenshot:

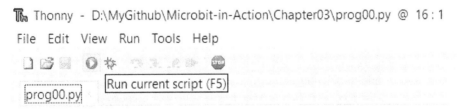

Figure 3.1 – Executing a Python script

You can also press the shortcut *F5* on the keyboard to run the program in the current editor window. We can also execute the program by selecting the options **Run | Run current script** from the main menu as shown in the following screenshot:

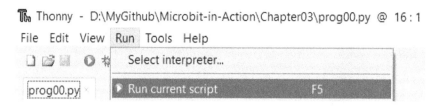

Figure 3.2 – Executing a Python script using the Run menu

The Thonny IDE executes the current program in the editor window and displays the result in the REPL (shell) window as shown in the following screenshot:

```
>>> print("Hello, World!")
 Hello, World!
>>>
```

Figure 3.3 – Output of a script execution

Congratulations! This is the very first program we wrote and executed from scratch. Note that the program is saved on the local computer, but it is running in the main memory of the BBC Micro:bit device connected to the computer. If we reboot the Micro:bit by disconnecting it and reconnecting it or by pressing *Ctrl + D* using the REPL, the program will not be there in the Micro:bit. We can also store the program with any name in the Micro:bit. And if we want the Micro:bit to run the program every time we power it up or reboot, then we can save the program in the Micro:bit with the name main.py. While practicing programming, I usually save the files on the local computer and run them on Micro:bit from the Thonny IDE. While deploying projects (such as a robot), I create a main.py file in the Micro:bit and save the code there.

Code comments

We can add comments to the code. Code comments are the parts of the program that are not executed. However, they provide additional information about the program. Comments can be thought of as basic documentation of the program. We can write a single-line comment by using # as follows:

```
# Sample program
# Author : Ashwin Pajankar

a = 1
b = 2.14
print(a + b)
print("My first script!!")
```

If you run the program, then the output will be the same as earlier. This is because the interpreter ignores the text in the comments. We can also have multiline comments as follows:

```
# Sample program
# Author : Ashwin Pajankar

'''Multiline comment line 1
line 2
'''

"""Another way of writing
multiline comments"""

a = 1
b = 2.14
print (a + b)
print ("My first script!!")
```

You must have noticed that the Thonny IDE highlights the different components of the program in different colors. This is one of the advantages of using IDEs for programming.

Let's practice a few more programs now.

Arithmetic, string, and logical operations

The following are examples of arithmetic, string, and logical operations:

```
a = 10
b = 2
c = 4
d = a + b / c
print (d)

str1 = "Test"
str2 = 'string'
str3 = str1 + ' ' + str2
print (str3)

b_val1 = True
```

```
b_val2 = False
print(b_val1 and b_val2)
print(b_val1 or b_val2)
print(not b_val1)
```

The output is as follows:

```
>>> %Run -c $EDITOR_CONTENT
10.5
Test string
False
True
False
```

Data type conversion

In Python, we can change the data type of variables. This is known as **Data Type Conversion**. Consider the following program:

```
str1 = "The value of Pi is "
pi = 22/7
print(str1 + pi)
```

The output is as follows:

```
>>> %Run -c $EDITOR_CONTENT
Traceback (most recent call last):
  File "<stdin>", line 3, in <module>
TypeError: can't convert 'float' object to str implicitly
```

This is because the variables str1 and pi are not of the same data type. Our goal is to print a string. So, it is logical to convert the variable pi into a string. The built-in str() method does the same. Let's rectify the program as follows:

```
str1 = "The value of Pi is "
pi = 22/7
print(str1 + str(pi))
```

The output is as follows:

```
>>> %Run -c $EDITOR_CONTENT
The value of Pi is 3.142857
```

Following is another example:

```
str1 = "3.14"
str2 = "1234"
pi = float(str1)
num = int(str2)
comp = complex(2, 3)
print(pi)
print(num)
print(comp)
```

The output is as follows:

```
>>> %Run -c $EDITOR_CONTENT
3.14
1234
(2+3j)
```

Handling user input

We can accept user input using the built-in method `input()`. The following is the sample code:

```
name = input("Enter your name :")
print("Welcome " + name)
```

The output will prompt you for input. After entering an input, it shows the following output:

```
>>> %Run -c $EDITOR_CONTENT
Enter your name: Ashwin
Welcome Ashwin
```

We will soon use this concept to learn other important concepts in programming. Next, we will learn about and experiment with conditional statements in Python.

Conditional statements

We can write conditional statements in Python. The conditional flow can be achieved with the `if` keyword. It is used in combination with a Boolean expression. If the expression returns `True`, then the block listed under `if` is executed. In Python, code blocks are denoted with indentation. In fact, indentation is the only way to create a code block (unlike { and } in C, C++, and Java). Following is an example of the usage of the `if` statement:

```
num = int(input("Enter a number :"))


if num > 10 :
    print("The entered number is greater than 10.")
```

Run the program and see the output. We can enhance our logic with the `else` clause as follows:

```
num = int(input("Enter a number :"))


if num > 10 :
    print("The entered number is greater than 10.")
else:
    print("The entered number is less than or equal to 10.")
```

We can also create an `elif` ladder as follows:

```
num = int(input("Enter a number :"))


if num > 10 :
    print("The entered number is greater than 10.")
elif num == 10:
    print("The entered number is equal to 10.")
else:
    print("The entered number is less than 10.")
```

This is how we can use an `elif` construct in Python.

Loops

Looping means repeating something. It is a very common programming construct. We can use it to perform repetitive actions in our programs. Let's start with the `while` construct. This construct evaluates a Boolean expression in every loop, and as long as the expression returns `True`, the code block under while is run repeatedly. When the Boolean expression returns False, the loop terminates. Following is a simple example:

```
i = 0
while i < 10:
    print(i)
    i = i + 1
```

The output is as follows:

```
>>> %Run -c $EDITOR_CONTENT
0
1
2
3
4
5
6
7
8
9
```

Sometimes, we may wish to run the loop forever. In such cases, we can use 1 or True in place of the Boolean expression in the while construct as follows:

```
#while 1
while True:
    print("Processing...")
```

In order to exit the loop, press *Ctrl + C* in the shell. Following is the output:

```
Processing...
Processing...
Processing...
Processing...
Traceback (most recent call last):
  File "<stdin>", line 3, in <module>
KeyboardInterrupt:
```

Such loops are known as infinite loops and can only be interrupted by manually interrupting the program under execution. If we make a mistake in the Boolean expression so that it never returns `False`, then our loop turns into an infinite loop. So, we have to be a bit careful while writing loops.

We can also have `for` loops. In this section, we will learn how to create `for` loops with the keywords `for` and `in` and the built-in method `range()`. The method `range()` creates a range from the given arguments. Following are a few possible cases of usage. You can figure out the functionality of the method `range()` from the arguments passed:

```
for i in range(5):
    print(i)
print('------------')


for i in range(5, 10):
    print(i)
print('-----------')


for i in range(0, 10, 2):
    print(i)
print('------------')
```

The output is as follows:

```
>>> %Run -c $EDITOR_CONTENT
0
1
2
3
4
------------
5
6
7
8
9
------------
0
2
4
```

```
6
8
- - - - - - - - - - - -
```

In the first case, we are iterating from 0 to 4. In the second case, we are iterating from 5 to 9. The second argument passed to the call of `range()` is non-inclusive. In both cases, the size of the step is 1. In the final case, we are iterating from 0 to 9 in increments of 2.

To have a better understanding of the mechanism, change the values of the arguments passed to the call of the method `range()`. As an exercise, print all the odd numbers between 1 to the number accepted as user input with the method `input()`.

In the next section, we will see the applications of the concepts we just learned. We will use the programming constructs we learned to compute prime numbers, Fibonacci series, and factorials.

Computing prime numbers, factorials, and Fibonacci series

We can find out if a number is prime by checking its divisibility with all the numbers less than it. If a number is only divisible by 1 and itself, then it's a prime number. We can compute prime numbers in a given range using the `for` loop twice as follows:

```python
lower = 2
upper = 100
for num in range (lower, upper+1):
    for i in range(2, num):
        if(num % i) == 0:
            break
    else:
        print(num)
```

The output is as follows:

```
>>> %Run -c $EDITOR_CONTENT
2
3
5
7
...
97
```

We can also compute the factorial of a number. The factorial of 0 and 1 is 1. For the rest of the integers, the multiplication of all the integers from 1 to that integer is factorial. Let's write a simple program that computes the factorial of a given number as follows:

```
num = int(input("Please enter an integer: "))
fact = 1
if num == 0:
    print("Factorial of 0 is 1.")
else:
    for i in range (1, num+1):
        fact = fact * i
    print("The factorial of {0} is {1}.".format(num, fact))
```

In the preceding program, we are already familiar with all the code except the method `format()`. It formats the specified values and replaces them with the string's placeholder. Following is the output of a few sample runs:

```
>>> %Run -c $EDITOR_CONTENT
Please enter an integer: 0
Factorial of 0 is 1.
>>> %Run -c $EDITOR_CONTENT
Please enter an integer: 1
The factorial of 1 is 1.
>>> %Run -c $EDITOR_CONTENT
Please enter an integer: 5
The factorial of 5 is 120.
>>> %Run -c $EDITOR_CONTENT
Please enter an integer: 10
The factorial of 10 is 3628800.
>>> %Run -c $EDITOR_CONTENT
Please enter an integer: 20
The factorial of 20 is 2432902008176640000.
```

Each number is the sum of the two preceding ones in the Fibonacci sequence. It begins with 0 and 1. The third number is 0 + 1 = 2. The fourth number is 1 + 2 = 3. The fifth number is 2 + 3 = 5. And so on, we can calculate as many numbers in this series as we desire. The following program demonstrates 20 numbers in this sequence. We are already familiar with all the syntax used in the code:

```
a = 0
b = 1
i = 0
while i < 20:
    print(a)
    c = a + b
    a = b
    b = c
    i = i + 1
```

The output is as follows:

```
>>> %Run -c $EDITOR_CONTENT
0
1
1
2
3
5
...
610
987
1597
2584
4181
```

So, this is how we can use loops to compute a list of prime numbers, Fibonacci series, and factorials. Practicing with these examples has now got us acquainted with the syntax of MicroPython and the Thonny IDE.

Summary

In this chapter, we learned the basics of Python syntax. The advantage of this chapter is that if we wish to use this knowledge elsewhere for Python programming, we can do so. As an exercise for this chapter, run all these programs using the reference Python 3 interpreter, CPython, provided by Python Software Foundation (`www.python.org`). All the knowledge we gained in this chapter will be employed when we start tinkering with the hardware of Micro:bit down the line.

In the next chapter, we will continue our journey and explore the advanced functionalities offered by Python.

Further reading

MicroPython's home page has a lot of documentation hosted at `https://docs.micropython.org/en/latest/`. It is worth exploring it. Also, if you are interested, you can visit the documentation of Python 3 hosted at `https://docs.python.org/3/`. The MicroPython port specific to Micro:bit is documented at `https://microbit-micropython.readthedocs.io/en/v2-docs/`.

4

Advanced Python

In the previous chapter, we got acquainted with the basics of Python programming syntax. While there was nothing specific to MicroPython and Micro:bit, the knowledge we learned is like a Swiss Army knife that can be used in every Python programming project.

In this chapter, we will continue our journey of exploring Python syntax. In the final part of this chapter, we will also explore some MicroPython- and Micro:bit-specific functionality. We will explore the following list of topics together:

- Lists, tuples, and dictionaries

- Functions

- Recursion

- Object-oriented programming with Python

- Getting help for built-in modules

- Retrieving system properties with code

Let's explore advanced concepts in programming.

Technical requirements

This chapter does not require any additional hardware. A Micro:bit board, a computer, and a micro USB cable are enough to follow the demonstrations mentioned in the chapter.

Lists, tuples, and dictionaries

Python comes with a lot of built-in data structures. In this section, we will explore a few of them. We will use the shell (REPL) for this. Let's get started with lists. Lists can store more than one item. They are defined with square brackets, and a comma separates their elements. Lists are mutable. This means that we can change the items in lists. They also allow duplicates. Open the REPL of the IDE of your choice and start following the examples:

```
>>> office_suites = ["Microsoft Office", "LibreOffice", "Apache
OpenOffice", "FreeOffice", "WPS Office", "Polaris Office",
"StarOffice", "NeoOffice", "Calligra Suite", "OnlyOffice"]
```

We can see the values in the dictionary on the REPL shell console by typing in the name of the list as follows:

```
>>> office_suites
['Microsoft Office', 'LibreOffice', 'Apache OpenOffice',
'FreeOffice', 'WPS Office', 'Polaris Office', 'StarOffice',
'NeoOffice', 'Calligra Suite', 'OnlyOffice']
```

We can also use the print() method as follows:

```
>>> print(office_suites)
['Microsoft Office', 'LibreOffice', 'Apache OpenOffice',
'FreeOffice', 'WPS Office', 'Polaris Office', 'StarOffice',
'NeoOffice', 'Calligra Suite', 'OnlyOffice']
```

We can also access the individual elements of a list using the C-style indexing scheme. The index of a list with n number of items starts at 0 and ends at n-1:

```
>>> office_suites[0]
'Microsoft Office'
>>> office_suites[1]
'LibreOffice'
>>> office_suites[6]
'StarOffice'
```

We can use the built-in len() method as follows to find out the length of the list:

```
>>> len(office_suites)
10
```

We can access the final and the penultimate elements of the list as follows:

```
>>> office_suites[-1]
'OnlyOffice'
>>> office_suites[-2]
'Calligra Suite'
```

If we try to access a nonexistent element, we get the following error:

```
>>> office_suites[13]
Traceback (most recent call last):
  File "<stdin>", line 1, in <module>
IndexError: list index out of range
```

We can also create lists with the following alternative syntax:

```
>>> office_suites = list (('Microsoft Office', 'LibreOffice',
'Apache OpenOffice', 'FreeOffice', 'WPS Office', 'Polaris
Office', 'StarOffice', 'NeoOffice', 'Calligra Suite',
'OnlyOffice'))
>>> office_suites
['Microsoft Office', 'LibreOffice', 'Apache OpenOffice',
'FreeOffice', 'WPS Office', 'Polaris Office', 'StarOffice',
'NeoOffice', 'Calligra Suite', 'OnlyOffice']
```

We can access a range of elements from the list:

```
>>> office_suites[2:5]
['Apache OpenOffice', 'FreeOffice', 'WPS Office']
```

We can also access all the elements after a particular index:

```
>>> office_suites[2:]
['Apache OpenOffice', 'FreeOffice', 'WPS Office', 'Polaris
Office', 'StarOffice', 'NeoOffice', 'Calligra Suite',
'OnlyOffice']
```

We can access all the elements before an index:

```
>>> office_suites[:2]
['Microsoft Office', 'LibreOffice']
```

We can change an element at a specific index as follows:

```
>>> office_suites[0] = 'GNOME Office'
```

We can insert an item at a particular index, and the items after that index will be automatically shifted:

```
>>> office_suites.insert(2, 'SoftMaker Office')
```

We can append an item at the end of the list:

```
>>> office_suites.append('Microsoft Office')
```

We can remove an item of a particular value from the list as follows:

```
>>> office_suites.remove('Microsoft Office')
```

We can iterate over a list using a `for` loop as follows. Save the following code as a Python program and run it:

```
office_suites = list (('Microsoft Office', 'LibreOffice',
                      'Apache OpenOffice', 'FreeOffice',
                      'WPS Office', 'Polaris Office',
                      'StarOffice', 'NeoOffice',
                      'Calligra Suite', 'OnlyOffice'))

for i in office_suites:
    print(i)
```

Similarly, we can employ a `while` loop for the same purpose as follows:

```
office_suites = list (('Microsoft Office', 'LibreOffice',
                      'Apache OpenOffice', 'FreeOffice',
                      'WPS Office', 'Polaris Office',
                      'StarOffice', 'NeoOffice',
                      'Calligra Suite', 'OnlyOffice'))

i = 0

while i < len(office_suites):
    print(office_suites[i])
    i = i + 1
```

Let's have a brief look at the concept of tuples. Tuples are immutable in nature. This means that once created, they cannot be changed. We can use tuples to store constant arrays of values. The following is an example of tuples:

```
>>> languages = ('C', 'C++', 'Python')
```

Just like lists, we can access languages using indices:

```
>>> languages[0]
'C'
>>> languages[-1]
'Python'
```

We can demonstrate the immutability of tuples by trying to change a value as follows:

```
>>> languages[1] = 'Java'
```

We get the following error:

```
Traceback (most recent call last):
  File "<stdin>", line 1, in <module>
TypeError: 'tuple' object doesn't support item assignment
```

Let's understand the concept of dictionaries. Dictionaries are ordered, mutable (changeable), and do not allow duplicates. Items are stored in dictionaries in pairs of **keys** and **values**. The following is an example of a simple dictionary:

```
>>> dev_env = {"language" : "C", "compilers": ['GCC', 'llvm',
'MinGW-w64', 'VC++']}
>>> dev_env
{'compilers': ['GCC', 'llvm', 'MinGW-w64', 'VC++'], 'language':
'C'}
```

We can access a value using its key as follows:

```
>>> dev_env['language']
'C'
>>> dev_env['compilers']
['GCC', 'llvm', 'MinGW-w64', 'VC++']
```

We can also retrieve all the keys and all the values, respectively, as follows:

```
>>> dev_env.keys()
dict_keys(['compilers', 'language'])
>>> dev_env.values()
dict_values([['GCC', 'llvm', 'MinGW-w64', 'VC++'], 'C'])
```

We can update/change a value as follows:

```
>>> dev_env.update({"language": "C++"})
>>> dev_env
{'compilers': ['GCC', 'llvm', 'MinGW-w64', 'VC++'], 'language':
'C++'}
```

We can also add a pair of keys and values as follows:

```
>>> dev_env['OS'] = ["UNIX", "FreeBSD", "Linux", "Windows 11"]
>>> dev_env
{'compilers': ['GCC', 'llvm', 'MinGW-w64', 'VC++'], 'OS':
['UNIX', 'FreeBSD', 'Linux', 'Windows 11'], 'language': 'C++'}
```

This is how we can work with lists, tuples, and dictionaries. Next, let's look at functions.

Functions

Let's understand the concept of functions. They are also called subroutines. They are a common programming practice. If a block of code is too long and repetitively used in the program, then we write that block separately from the other code and assign it a name. We call the block using the assigned name wherever needed. Let's see an example:

```
def message():
    name = input("What is your name, My liege : ")
    print("Ashwin is your humble servant, My liege " + name)

print("First function call...")
message()
print("Second function call...")
message()
```

In this example, we have defined our own function and named it message(). The definition of the function begins with the def keyword. The output is as follows:

```
>>> %Run -c $EDITOR_CONTENT
First function call...
What is your name, My liege : Henry V
Ashwin is your humble servant, My liege Henry V
Second function call...
What is your name, My liege : Caesar
Ashwin is your humble servant, My liege Caesar
```

This is a very simple example. We can also define a function with parameters, and we can pass arguments to those parameters in the function call as follows:

```
def printHello( first_name ):
    print("Hello {0}".format(first_name))

def printHello1( first_name, last_name ):
    print("Hello {0}, {1}".format(first_name, last_name))

def printHello2( first_name, msg='Good Morning!'):
    print("Hello {0}, {1}".format(first_name, msg))


printHello('Ashwin')
printHello(first_name='Thor')
printHello1('Ashwin', 'Pajankar')
printHello1(first_name='Ashwin', last_name='Pajankar')
printHello1(last_name='Pajankar', first_name='Ashwin')
printHello1('Thor', 'Odinson')
printHello2('Ashwin', 'Good Evening!')
printHello2('Thor')
```

The first definition is a very basic example. In the second definition, we are defining a function with multiple parameters. In the third definition, we are defining a function with the default argument.

We have tried different ways to call the functions. Sometimes, we have mentioned the names of the parameters. We have also changed the order of the parameters in the function call. Also, we have tried the function call with the default arguments. The following is the output:

```
>>> %Run -c $EDITOR_CONTENT
```

```
Hello Ashwin
Hello Thor
Hello Ashwin, Pajankar
Hello Ashwin, Pajankar
Hello Ashwin, Pajankar
Hello Thor, Odinson
Hello Ashwin, Good Evening!
Hello Thor, Good Morning!
```

Perhaps it is a good idea to change the arguments and their order in the function call to understand this concept better. Try that yourself as an exercise.

A function can also be defined as returning a value or multiple values. We can store these values in variables. The following is a simple example:

```
def square(a):
    b = a**2
    return b

def power(a, b):
    return a**b

def multiret(a, b):
    a += 1
    b += 1
    return a, b

num = square(3)
print(num)

num = power(3, 3)
print(num)

x, y = multiret(3, 5)
print("{0}, {1}".format(x, y))
```

The output is as follows:

```
>>> %Run -c $EDITOR_CONTENT
9
27
4, 6
```

We can also define a function to accept an arbitrary number of arguments as follows:

```
def team( *members ):
    print("\nMy team has following members :")
    for name in members:
        print( name )


team('Ashwin', 'Jane', 'Thor', 'Tony')
team('Ashwin', 'Jane', 'Thor')
```

\n in print () refers to a newline. It prints the entire string on the next line. The output is as follows:

```
>>> %Run -c $EDITOR_CONTENT

My team has following members :
Ashwin
Jane
Thor
Tony

My team has following members :
Ashwin
Jane
Thor
```

Let's learn about global and local variables now. A variable can be created outside and inside a function. The variable created inside a function is known as a local variable, and its scope is limited to that function, whereas a global variable can be accessed anywhere. The following program demonstrates a global and a local variable:

```
x = "Python"
```

```
def function01():
    x = "C++"
    print(x + " is the best language.")


function01()
print(x + " is the best language.")
```

In the preceding program, the x variable defined outside `function01()` is a **global variable**. The x variable defined inside `function01()` is a **local variable** of that function. The output is as follows:

```
>>> %Run -c $EDITOR_CONTENT
C++ is the best language.
Python is the best language.
```

We can also access the global variable inside a function as follows:

```
x = "Python"

def function01():
    global x
    print(x + " is the best language.")


function01()
```

Run the program and check the output. In the following section, we will study another important concept, recursion.

Recursion

In Python, just like many other programming languages, a defined function can call itself. This is known as **recursion**. A recursive pattern has two important components. The first is the termination condition, which defines when to terminate the recursion. In the absence of this or if it is erroneous (never meets the termination criteria), the function will call itself an infinite number of times. This is called infinite recursion. The other part of the recursion pattern is recursive calls. There can be one or more of these. The following program demonstrates simple recursion:

```
def print_message(string, count):
    if (count > 0):
        print(string)
```

```
        print_message(string, count - 1)

print_message("Hello, World!", 5)
```

In the preceding program, we use recursion in place of a `for` or `while` loop to print a message repetitively.

Let's write a program for computing a factorial as follows:

```
def factorial(number):
    if number < 0:
        print("The number must be greater than zero.")
        return -1
    elif ((number - int(number)) > 0):
        print("The number must be a positive integer.")
        return -1
    elif (number == 0) or (number == 1):
        return 1
    else:
        return number * factorial(number-1)

factorial(-1)
factorial(1.1)
print(factorial(0))
print(factorial(1))
print(factorial(5))
```

The output is as follows:

```
The number must be greater than zero.
The number must be a positive integer.
1
1
120
```

The following program employs recursion to return the number in the Fibonacci sequence at the given index:

```
def fibonacci(number):
    if number <= 1:
        return number
```

```
    else:
        return fibonacci(number-1) + fibonacci(number-2)

print(fibonacci(3))
```

Indirect recursion

The type of recursion we have seen in the preceding three examples is known as direct recursion because the recursive function calls itself. We can write our program with two or more functions calling each other, creating something we call **indirect recursion**. The following is an example of a *ping-pong mechanism* demonstrated with indirect recursion:

```
def ping(i):
    if i > 0:
        print("Ping - " + str(i))
        return pong(i-1)

def pong(i):
    if i > 0:
        print("Pong - " + str(i))
        return ping(i-1)

ping(10)
```

The output is as follows:

```
Ping - 10
Pong - 9
Ping - 8
Pong - 7
Ping - 6
Pong - 5
Ping - 4
Pong - 3
Ping - 2
Pong - 1
```

This is how we can work with recursion. In the next section, we will briefly have a look at the concepts of the object-oriented programming paradigm in Python.

Object-oriented programming with Python

This is one of the most important sections in the chapter. The **object-oriented programming** paradigm revolves around the concept of **objects**. Objects contain data (in the form of **properties**) and functions (in the form of **procedures**, known as **methods** in Python programming). We will use a lot of these concepts in our projects. Let's create an integer variable as follows:

```
>>> a = 10
```

We can use the built-in type() method to see the data type of the variable as follows:

```
>>> type(a)
<class 'int'>
```

We can do this for the other types of values, too, as follows:

```
>>> type("Hello, World!")
<class 'str'>
>>> type(3.14)
<class 'float'>
```

This means that everything is an object in Python. An object is a variable of a class data type. These classes could be built-in library classes or user-defined classes. Let's see a simple example of a user-defined class as follows:

```
class fruit:

    def __init__(self, str):
        self.name = str

    def printName(self):
        print(self.name)


f1 = fruit('Mango')
print(type(f1))
f1.printName()
```

The output is as follows:

```
>>> %Run -c $EDITOR_CONTENT
<class 'fruit'>
Mango
```

We are using the `class` keyword to define a user-defined class. The constructor or initializer is defined with `__init__` and is called implicitly when we create an object of the class. The variable name is a property of the class. The function defined within a class is known as a **method** or a **class method**. `printName()` is one such class method. For practice, create your own classes to represent real-life concepts. For example, a class corresponding to a computer can have properties, such as processor name, processor speed, RAM type, RAM size, graphics card model, and so on. It can also have methods such as `powerOff()`, `powerOn()`, and `reset()`.

We can also use the classes created in one program in another program. Basically, a Python code file is also known as a module. And we can use the `import` keyword to use such modules in other Python programs. We will use plenty of such standard Python and MicroPython library modules for the demonstrations in this book.

The **object-oriented programming paradigm** is a vast topic, and it is not feasible to cover it entirely in a single chapter. So, we will explore only the features that are relevant to programming a Micro:bit.

Exploring the random module

Let's generate random numbers with a built-in `random()` module. Type the following code in to the REPL shell:

```
>>> import random
```

This imports the module to our shell (or Python program). The module has many methods. We will see a few important ones. We can obtain any random float between `0.0` and `0.1` with the `random()` method as follows:

```
>>> random.random()
0.2952699
```

We can obtain a random integer in the given range with the `randint()` method as follows:

```
>>> random.randint(2, 5)
4
```

We can obtain a random number (a floating point number) in the given range with the `uniform()` method as follows:

```
>>> random.uniform(2, 5)
3.330382
```

In the case that we do not wish to use the name of the module with the method, we have to import in the module the following way:

```
>>> from random import *
```

The following is a program that demonstrates this style of coding:

```
from random import *
print(random())
print(randint(2, 5))
print(uniform(2, 5))
```

We will use both styles of coding in this book.

In the next section, we will explore how to get help for the built-in library modules in Python.

Getting help for built-in modules

We can get help for the built-in modules with the following command in the REPL shell:

```
>>> help('modules')
```

This shows the list of all the available modules as follows:

```
>>> help('modules')
__main__            machine            os              uerrno
antigravity         math               radio           urandom
audio               microbit           speech          ustruct
builtins            micropython        this            usys
gc                  music              uarray          utime
love                neopixel           ucollections
Plus any modules on the filesystem
```

Figure 4.1 – Built-in modules

Note that the output is interpreter-specific. The preceding screenshot (*Figure 4.1*) shows the output specific to the MicroPython interpreter implemented for the Micro:bit. If we run the same statement on any other interpreter of Python (for example, CPython, the reference implementation interpreter of Python by the Python Software Foundation), we will see a different list.

We can mention the name of any built-in module as an argument to the `help()` method to see the details of the module as follows:

```
>>> help('os')
 object os is of type str
    decode -- <function>
    encode -- <function>
    find -- <function>
    rfind -- <function>
    index -- <function>
    rindex -- <function>
    join -- <function>
```

Figure 4.2 – Obtaining help

All the code we have practiced before now in this chapter and the previous one can be executed with the standard reference implementation of Python 3 (www.python.com). As an exercise, execute all the statements and programs on the IDLE shell interpreter.

Retrieving system properties with code

Now, let us see some Micro:bit-specific MicroPython code. This code is only run on MicroPython running on the Micro:bit. It won't run on the standard Python 3 interpreter. Let's see the properties of our Micro:bit device as follows:

```
# micro:bit specific MicroPython code
import machine
print("Unique ID : " + str(machine.unique_id()))
print(str(machine.freq()/1000000) + " MHz")
```

The program prints the unique ID and the frequency of our Micro:bit board. The following is the output:

```
>>> %Run -c $EDITOR_CONTENT
Unique ID : b'$\xd1\xb4[\xe2a?\xd5'
64.0 MHz
```

This is how we can fetch and display system properties.

Summary

In this chapter, we explored the advanced features of the Python programming language. The final program was specific to MicroPython running on a Micro:bit. From the next chapter onward, all the code demonstrations will be specific to MicroPython running on a Micro:bit, and they will not run on any Python 3 implementation.

After completing this chapter, you will understand the power of the Python programming language. It is simple yet concise and powers devices such as the Micro:bit and other microcontrollers (such as the Raspberry Pi Pico W and ESP32) in the form of MicroPython. Anyone can learn to work with Python and MicroPython. The following is a photograph of my neighbors in my village working on a Micro:bit connected with NeoPixel Ring (we will fully explore this in *Chapter 11, Working with NeoPixels and a MAX7219 Display*) for their school project:

Figure 4.3 – Schoolkids working on a NeoPixel with a Micro:bit

From the next chapter onward, we will start creating projects with hardware programming. In the next chapter, we will explore how to program a built-in LED matrix and push buttons in depth.

Further reading

The basics we have learned about up to now are available in the format of video tutorials on YouTube on my channel at https://www.youtube.com/watch?v=PITLKocdY14&list=PLiHals-EL3vhJTNPXOkjsWDes9cMZRo1D. I have used Micro:bit V1 for recording the tutorials. However, they are not outdated. All the videos are consolidated into a single lecture at https://www.youtube.com/watch?v=YcHZj_B6X18. It will be worth exploring these tutorials.

Part 2:
Programming Hardware
with MicroPython

This section is dedicated to the programming concepts of MicroPython. You will get started with the programming hardware of the BBC Micro:bit with MicroPython. This section aims to make you comfortable with the programming of built-in external LEDs and buttons. The section also explores the programming of buzzers and stepper motors.

This section has the following chapters:

- *Chapter 5, Built-in LED Matrix Display and Push Buttons*
- *Chapter 6, Interfacing External LEDs*
- *Chapter 7, Programming External Push Buttons, Buzzers, and Stepper Motors*

5

Built-in LED Matrix
Display and Push Buttons

In the previous chapter, we got acquainted with advanced concepts in Python programming and syntax. We also wrote and executed a Micro:bit-specific MicroPython program. Going forward, we will use those concepts in all of our demonstrations with Micro:bit.

The previous two chapters mostly served as an introduction to the syntax of Python. We did not use any Micro:bit-specific hardware features and libraries. From this chapter onward, all the programs will be specific to the MicroPython flavor for the Micro:bit as all of them will employ and demonstrate the hardware capabilities of MicroPython on Micro:bit. They will not execute on the other implementations of Python such as Python 3 or any MicroPython implementation for other devices such as Raspberry Pi Pico W and ESP8266 or ESP32.

This chapter explores two very important hardware features of the Micro:bit and demonstrates them with MicroPython. The built-in components include a 5x5 LED matrix display and a pair of push buttons. We will explore the following list of topics together:

- Built-in programmable 5x5 LED matrix

- Images and animations

- Working with built-in push buttons

Let's explore the built-in matrix display and push buttons.

Technical requirements

This chapter does not require any additional hardware. A Micro:bit board, a computer, and a micro USB cable are enough to follow the demonstrations in the chapter.

Built-in programmable 5x5 LED matrix

LED stands for **light-emitting diode**. Just like a normal diode, an LED allows the current to flow only in one direction. An LED has two pins (or legs): an *anode* and a *cathode*. Just as with any other diode, we must connect the anode of an LED to a positive pin and the cathode to the ground or negative pin of the power source. Then, the LED will allow the current to flow through it. However, if we connect the anode pin to the ground/negative pin and the cathode to the positive pin of the source, the LED will not let the current pass through it. When current passes through LEDs, they glow. We will study LEDs and their programming in detail in this and the next chapter. This chapter focuses on the built-in programmable 5x5 LED matrix of the Micro:bit board and the next chapter focuses on external discrete LEDs. *Figure 5.1* shows an LED matrix. It has 5 rows and 5 columns, totaling 25 LEDs:

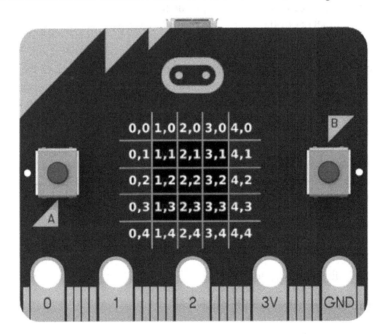

Figure 5.1 – Built-in programmable 5x5 LED matrix (Courtesy: https://commons. wikimedia.org/wiki/File:Micro_bit_position_des_DEL.png)

Figure 5.1 shows their positional indices, which are used to refer to them in the MicroPython programs we will write and execute. The LEDs glow with in red. Let's get started with the programming of the LED matrix.

Displaying characters and text

Let's see how to display characters, numbers, and strings. Here is a sample program:

```
from microbit import *
if not display.is_on():
    display.on()
else:
    print("Display is already on...")
a = 'A'
display.show(a)
sleep(1000)
a = 'BBC'
display.show(a)
sleep(1000)
a = 'MicroPython'
display.show(a, delay=750, clear=True)
sleep(1000)
a = 'Micro:bit'
display.show(a, delay=750, loop=True)
sleep(1000)
```

The library mentioned in the first line of the program, microbit, is a MicroPython library specific to Micro:bit. The next if statement checks if display.is_on() returns True. If not, then it calls display.on() to turn the built-in display on. Then, the display.show() method shows a single character on display. The sleep() method suspends (or waits) the current thread's execution for a given number of milliseconds. 1 second equals 1,000 milliseconds. The display continues to show the given character while the Micro:bit sleeps. Then, we show a string with the show() method. It shows all the characters in sequence and the final character in the string while the Micro:bit sleeps. Next, we are showing a string with a custom delay between characters defined by the 750 argument to the delay parameter. Because we passed the True argument to the clear parameter, the Micro:bit clears the display after showing the final character in the sequence. As a result, while it sleeps for a second, the display is blank. We are showing the final string in a loop. Once we enable it with the parameter loop, the next statements will not be executed. Run the program and see the code in action.

We can use the **exception handling** feature in Python programming to interrupt the forever loop by pressing *Ctrl + C* on the keyboard, as follows:

```
from microbit import *
if not display.is_on():
```

```
        display.on()
else:
        print("Display is already on...")
a = 'A'
display.show(a)
sleep(1000)
a = 'BBC'
display.show(a)
sleep(1000)
a = 'MicroPython'
display.show(a, delay=750, clear=True)
sleep(1000)
try:
        a = 'Micro:bit'
        display.show(a, delay=750, loop=True)
        sleep(1000)
except KeyboardInterrupt as e:
        print("Interrupted by the user...")
print("Program ended..")
```

Run the program and see the exception handling in action. In the final block of code, we are including the code in the try block where we suspect we would encounter an exception (in this case, the keyboard interrupt). We can then catch the exception in the except block and process it. However, there is one major catch. We can use this for debugging only when Micro:bit is connected to a computer with a keyboard.

We can also display numbers as follows. I have just modified the preceding program:

```
from microbit import *
if not display.is_on():
        display.on()
else:
        print("Display is already on...")
a = 1
display.show(a)
sleep(1000)
a = 1234
display.show(a)
```

```
sleep(1000)
a = 3.14
display.show(a, delay=750, clear=True)
sleep(1000)
try:
    a = 3.1416
    display.show(a, delay=750, loop=True)
    sleep(1000)
except KeyboardInterrupt as e:
    print("Interrupted by the user...")
print("Program ended..")
```

This LED display uses the digital GPIO pins of Micro:bit. We can turn off the display with the display.off() statement and use those GPIO pins. We will see this in detail in the next chapter.

Scrolling text on the display

We can also scroll text and numbers on the display as follows:

```
from microbit import *
if not display.is_on():
    display.on()
else:
    print("Display is already on...")
display.scroll(3.14)
display.scroll("Micro:bit", delay=300, monospace=True,
loop=True)
```

Here, we use the method scroll() to scroll the passed argument (a number or a string). We can pass arguments for the parameters delay (decides the speed of scrolling), loop, and monospace (characters and numbers consume all the five columns for their width).

Working with the individual LEDs

We can also write the code that addresses the individual LEDs. We can refer to *Figure 5.1* for the indexing scheme of LEDs. The individual LEDs are capable of emitting 10 different intensities of the red light. The intensity values range from 0 (lowest) to 9 (highest). Let's see a simple demonstration of that:

```
from microbit import *
if not display.is_on():
```

```
        display.on()
    else:
        print("Display is already on...")
    display.clear()
    k = 0
    for i in [0, 1]:
        for j in range(0, 5, 1):
            print (j, i, k)
            display.set_pixel(j, i, k)
            k = k + 1
```

The method display.clear() clears the display by setting all the LEDs to the intensity of 0 which means **off**. We are calling it in the beginning as we do not wish the output of the execution of the preceding program to interfere with the output of the execution of the current program. Then, we use a nested for loop and set_pixel() to set the intensities of the pixels in the first two rows from 0 to 9. The method set_pixel() accepts the column number, the row number, and the intensity as arguments. Run the program and see the output.

We can also use the method get_pixel() to fetch the current intensity of a LED. Add the following two lines at the end of the program and re-run it:

```
    print(display.get_pixel(0, 0))
    print(display.get_pixel(0, 1))
```

The output is as follows:

```
>>> %Run -c $EDITOR_CONTENT
Display is already on...
0
5
```

We can also create a simple animation as follows:

```
from microbit import *
if not display.is_on():
    display.on()
else:
    print("Display is already on...")


display.clear()
```

```
values = [0, 1, 2, 3, 4]
print (values)
try:
    while True:
        for i in range(0, 5, 1):
            display.set_pixel(i, 0, values[i])
            values[i] = (values[i] + 1) % 10
            print(values)
        sleep(100)
except KeyboardInterrupt as e:
    print("Interrupted by the user...")
```

The first line of the LED display shows the list of intensity values scrolling in the REPL shell. Similarly, we can create 2D animation utilizing all the LEDs, as follows:

```
from microbit import *
if not display.is_on():
    display.on()
else:
    print("Display is already on...")

display.clear()

values = [[0, 1, 2, 3, 4],
          [1, 2, 3, 4, 5],
          [2, 3, 4, 5, 6],
          [3, 4, 5, 6, 7],
          [4, 5, 6, 7, 8]]
print (values)

try:
    while True:
        for i in range(0, 5, 1):
            for j in range(0, 5, 1):
                display.set_pixel(i, j, values[i][j])
                values[i][j] = (values[i][j] + 1) % 10
```

```
                    print(values)
            sleep(100)
except KeyboardInterrupt as e:
    print("Interrupted by the user...")
```

Run the program and see the animation in action. This way, we can create a lot of custom animations. The possibilities are unlimited.

The next section explores displaying simple images and animations.

Images and animations

We can display a custom-made image with the show() method. Let's see a demonstration:

```
from microbit import *
if not display.is_on():
    display.on()
else:
    print("Display is already on...")
display.clear()
pattern0 = Image("00000:"
                 "11111:"
                 "22222:"
                 "33333:"
                 "44444")
display.show(pattern0)
```

Run the program to see the output. We can also define a pattern or an image in a single line in our program, as follows:

```
from microbit import *
if not display.is_on():
    display.on()
else:
    print("Display is already on...")
display.clear()
pattern0 = Image("00000:11111:22222:33333:44444")
display.show(pattern0)
```

We can also create a list of patterns and create a simple animation. We studied Python lists in the previous chapter and will use that concept in this demonstration:

```python
from microbit import *
if not display.is_on():
    display.on()
else:
    print("Display is already on...")
display.clear()
pattern0 = Image("00000:11111:22222:33333:44444")
pattern1 = Image("11111:22222:33333:44444:55555")
pattern2 = Image("22222:33333:44444:55555:66666")
pattern3 = Image("33333:44444:55555:66666:77777")
pattern4 = Image("44444:55555:66666:77777:88888")
pattern5 = Image("55555:66666:77777:88888:99999")
patterns = [pattern0, pattern1, pattern2,
            pattern3, pattern4, pattern5]
display.show(patterns, delay=100, loop=True)
```

In this program, we have created a list of patterns to be displayed. Then, we are passing that list to the show() method call to animate with a delay factor of 100 milliseconds between consecutive frames.

We also have plenty of built-in images. Here is a list of all of them:

```
Image.HEART
Image.HEART_SMALL
Image.HAPPY
Image.SMILE
Image.SAD
Image.CONFUSED
Image.ANGRY
Image.ASLEEP
Image.SURPRISED
Image.SILLY
Image.FABULOUS
Image.MEH
Image.YES
Image.NO
```

```
Image.CLOCK12, Image.CLOCK11, Image.CLOCK10, Image.CLOCK9,
Image.CLOCK8, Image.CLOCK7, Image.CLOCK6, Image.CLOCK5, Image.
CLOCK4, Image.CLOCK3, Image.CLOCK2, Image.CLOCK1
Image.ARROW_N, Image.ARROW_NE, Image.ARROW_E, Image.ARROW_SE,
Image.ARROW_S, Image.ARROW_SW, Image.ARROW_W, Image.ARROW_NW
Image.TRIANGLE
Image.TRIANGLE_LEFT
Image.CHESSBOARD
Image.DIAMOND
Image.DIAMOND_SMALL
Image.SQUARE
Image.SQUARE_SMALL
Image.RABBIT
Image.COW
Image.MUSIC_CROTCHET
Image.MUSIC_QUAVER
Image.MUSIC_QUAVERS
Image.PITCHFORK
Image.XMAS
Image.PACMAN
Image.TARGET
Image.TSHIRT
Image.ROLLERSKATE
Image.DUCK
Image.HOUSE
Image.TORTOISE
Image.BUTTERFLY
Image.STICKFIGURE
Image.GHOST
Image.SWORD
Image.GIRAFFE
Image.SKULL
Image.UMBRELLA
Image.SNAKE
Image.SCISSORS
```

We can use one of these built-in images, as follows:

```
from microbit import *
if not display.is_on():
    display.on()
else:
    print("Display is already on...")
display.clear()
display.show(Image.HEART)
```

We can create simple animations by creating a list of built-in images, as follows:

```
from microbit import *
if not display.is_on():
    display.on()
else:
    print("Display is already on...")
display.clear()
pattern = [Image.HEART, Image.XMAS, Image.PACMAN,
            Image.TARGET, Image.TSHIRT]
display.show(pattern)
```

The library also has a couple of built-in lists of images for animations. Those are Image.ALL_CLOCKS and Image.ALL_ARROWS. Let's use them in our programs, as follows:

```
from microbit import *
if not display.is_on():
    display.on()
else:
    print("Display is already on...")
display.clear()
display.show(Image.ALL_CLOCKS)
display.show(Image.ALL_ARROWS)
```

This is how we can use the built-in display. In the next section, we will explore the functionality offered by built-in push buttons.

Working with built-in push buttons

Figure 5.1 shows two built-in push buttons in Micro:bit. They are named **A** and **B**. They are connected to GPIO pins 5 and 11, respectively. The MicroPython implementation for Micro:bit comes with built-in methods to work with them. Let's see a simple example, as follows:

```
from microbit import *
try:
    while True:
        if button_a.is_pressed():
            print("Button A is pressed...")
        elif button_b.is_pressed():
            print("Button B is pressed...")
        else:
            print("No Button is pressed...")
        sleep(100)
except KeyboardInterrupt as e:
    print("Interrupted by the user...")
```

The is_pressed() method checks if an associated button is presently being pressed. It returns True if the button is pressed. We can also use the display to show the status of the button press, as follows:

```
from microbit import *
if not display.is_on():
    display.on()
else:
    print("Display is already on...")
display.clear()
try:
    while True:
        if button_a.is_pressed():
            display.show(Image.HAPPY)
        elif button_b.is_pressed():
            display.show(Image.SAD)
        else:
            display.show(Image.HEART)
        sleep(100)
```

```
except KeyboardInterrupt as e:
    print("Interrupted by the user...")
```

The was_pressed() method returns True at the release of the button, as follows:

```
from microbit import *
if not display.is_on():
    display.on()
else:
    print("Display is already on...")
display.clear()
try:
    while True:
        if button_a.was_pressed():
            display.show(Image.HAPPY)
        elif button_b.was_pressed():
            display.show(Image.SAD)
        else:
            display.show(Image.HEART)
        sleep(100)
except KeyboardInterrupt as e:
    print("Interrupted by the user...")
```

We can also see how many times a button is pressed, as follows:

```
from microbit import *
try:
    while True:
        sleep(5000)
        print("Button A has been pressed " + str(button_a.get_
presses()) + " times.")
        print("Button B has been pressed " + str(button_b.get_
presses()) + " times.")
except KeyboardInterrupt as e:
    print("Interrupted by the user...")
```

The Micro:bit sleeps for 5 seconds and counts button presses while it sleeps. We will use both buttons in the demonstrations in the coming chapters throughout the book.

Summary

We have learned how to work with built-in display and push buttons of Micro:bit. We also immersed ourselves in MicroPython programming that is specific to Micro:bit. Now, we have a fair understanding of how MicroPython interacts with the hardware features of Micro:bit. Like this one, the rest of the book's chapters will be full of hands-on programming with MicroPython using Micro:bit.

The next chapter explores the interfacing and programming of external LEDs and LED-based hardware with digital GPIO pins of Micro:bit and MicroPython.

Further reading

The following pages have more information on the programming of displays and buttons:

- https://microbit-micropython.readthedocs.io/en/latest/tutorials/images.html
- https://microbit-micropython.readthedocs.io/en/latest/tutorials/buttons.html

6
Interfacing External LEDs

In the previous chapter, we learned how to work with a built-in programmable LED matrix display and a pair of programmable built-in push buttons. We are now comfortable with using them in our projects.

This chapter explores how to connect with external discrete LEDs and products based on them. We will create exciting projects involving LEDs. We will also use our knowledge of working with built-in buttons to add interactivity to our projects. After that, we will start creating electronic circuits and program them with MicroPython.

The following is the list of topics that we will explore in this chapter:

- Breadboards and solderless circuits
- LEDs and their programming
- Chaser effect
- RGB LED
- Seven-segment display

Let's explore programming LEDs with MicroPython and a Micro:bit.

Technical requirements

Apart from the usual setup, the demonstrations in this chapter need the following components:

- An MB102 breadboard
- Jumper wires
- LEDs of different colors
- An LED bar graph
- A common cathode RGB LED

- A common anode RGB LED

- A common cathode seven-segment LED display

- A common anode seven-segment LED display

Breadboards and solderless circuits

We can create circuits with electrical and electronic components such as LEDs, wires, and buttons. We need to use the soldering method to connect them to make them work. However, there is another method we can use to avoid soldering altogether. This method is known as **breadboarding** or **prototyping**. We have to use a component known as a breadboard to avoid soldering other components. The following is an image of an MB102-type breadboard:

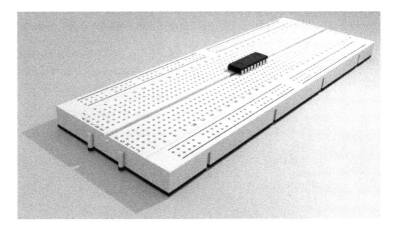

Figure 6.1 – MB102 breadboard (courtesy: https://commons.wikimedia.
org/wiki/File:Final_render_pic_on_breadboard.png)

A **Dual In-Line Package Integrated Circuit** (**DIP IC**) is mounted on the breadboard. Let's see how breadboards are useful. Have a look at the top view of the breadboard:

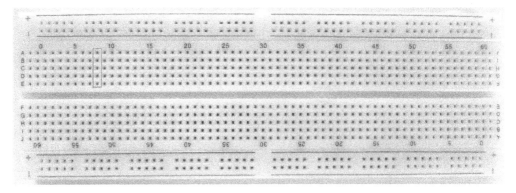

Figure 6.2 – The top view of an MB102 breadboard (courtesy: https://commons.
wikimedia.org/wiki/File:Bread_board_1480358_59_60_HDR_Enhancer.jpg)

The lines in *Figure 6.2* marked with red and blue are known as **bus lines**. All the points are connected with a single line. These bus lines are usually used to supply power to the breadboard. The breadboard has a central groove that divides it into two parts so that we can conveniently mount any electrical/ electronic component with two rows of male headers (refer to Figure 6.1). Furthermore, the two separated sections of the breadboard are grouped into collections of five points arranged vertically (refer to the green outline in Figure 6.2). All five points in a single collection are connected. This makes it easier for us to make multiple connections using a single pin of an electrical component. To connect the points to each other, as well as to various components, we can use jumper wires:

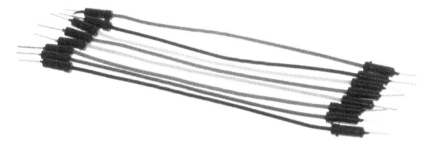

Figure 6.3 – Male-to-male header jumper wires (courtesy: https:// commons.wikimedia.org/wiki/File:A_few_Jumper_Wires.jpg)

The jumper wires shown in Figure 6.3 are **male-to-male** jumper wires. They are also available in **male- to-female** and **female-to-female** header types. We can also directly work with the edge connector of the Micro:bit using **crocodile clips**:

Figure 6.4 – Crocodile clips (courtesy: https://commons.wikimedia. org/wiki/File:Jumper_Wires_with_Crocodile_Clips.jpg)

These are the essential components for building solderless circuits. We will be using these components frequently to build circuits. This will help demonstrate the concepts we will learn about throughout this book. In the next section, we will explore circuit-making with LEDs.

LEDs and their programming

In the previous chapter, we learned about the definition of LEDs and worked with the built-in programmable 5 x 5 LED matrix. Let's see what a discrete LED looks like. Have a look at the following figure:

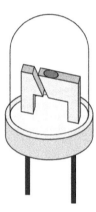

Figure 6.5 – An LED (courtesy: https://freesvg.org/1534357308)

An LED has got two connections – an anode (the longer leg in Figure 6.5) and a cathode (the shorter leg). An LED will glow when we connect the anode of the LED to positive voltage and the cathode to the ground. We know that the Micro:bit has a 3 V pin and a ground pin on its edge connector. We can use these pins with an LED for demonstration purposes, as follows:

Figure 6.6 – An LED connected to a Micro:bit

As we can see, it is difficult to understand the connections using photographs as we cannot see the wiring properly. That is why, for the circuit diagrams in the remainder of this book, we will use fritzing:

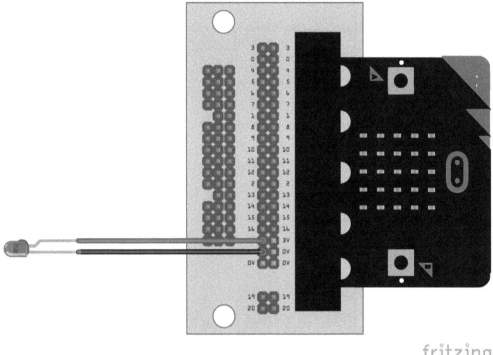

Figure 6.7 – A fritzing circuit

> **Note**
>
> We can also use Tinkercad (https://www.tinkercad.com) to create circuits. Tinkercad also simulates the circuit. It is a free online tool. As an exercise, explore Tinkercad.

As shown in Figure 6.7, connect the 3V pin of the Micro:bit to the anode of the LED and the GND to the cathode. The current will pass through the LED, and it will glow. We do not have to use any resistor in the circuit as we can also see the circuit schematics in fritzing:

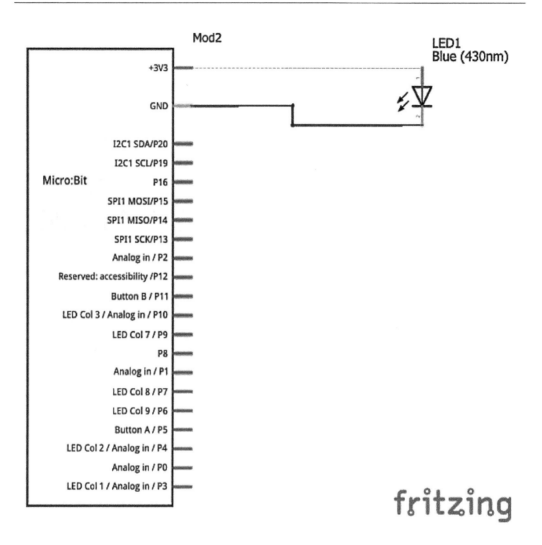

Figure 6.8 – A fritzing schematic for an LED connected to a Micro:bit

If we connect the cathode to the 3V pin and the anode to the GND pin, then the LED will not glow. Try this now!

Let's connect the anode of the LED to the digital **General-Purpose Input/Output (GPIO)** pin – that is, pin 0 – as shown in the following circuit diagram:

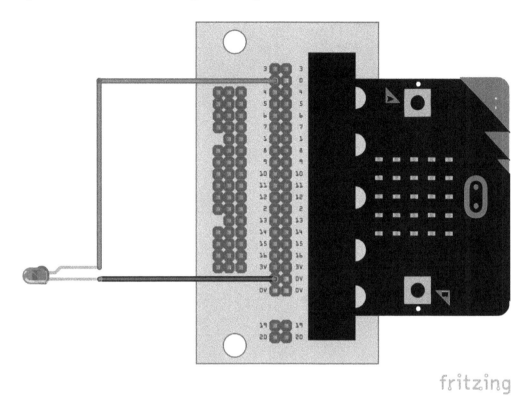

Figure 6.9 – An LED connected to pin 0 of a Micro:bit

The following is a schematic for this:

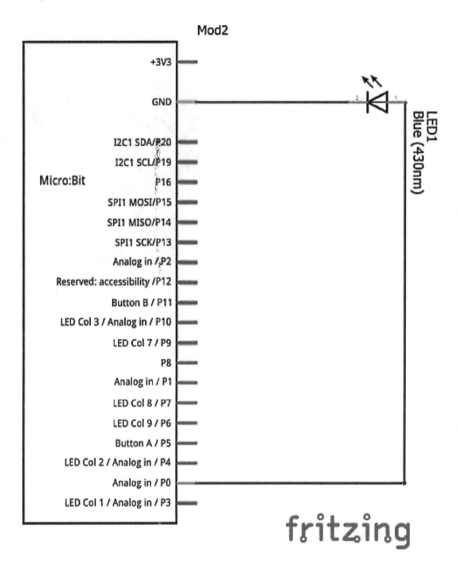

Figure 6.10 – Schematic of an LED connected to pin 0 of a Micro:bit

The LED will now glow automatically. We will have to write the following code to make it glow:

```
from microbit import *
pin0.write_digital(1)
```

We have to use the `write_digital()` method to send 3V or 0V to pin 0 (where 1 or 0 is used as an argument, respectively) of the Micro:bit.

Blinking an LED

We can even make the LED blink with the following code:

```
from microbit import *
try:
    while True:
        pin0.write_digital(1)
        sleep(500)
        pin0.write_digital(0)
        sleep(500)
except KeyboardInterrupt as e:
    print("Interrupted by the user...")
```

We are sending 3V and 0V to pin 0 of the Micro:bit for half a second in an alternate sequence.

SOS message

We know that three dots followed by three dashes, then followed by three dots (... --- ...), represents an SOS message in the standard *Morse code*. We can create a visual SOS message with the same circuit by uploading the following code to the Micro:bit:

```
from microbit import *
try:
    while True:
        pin0.write_digital(1)
        sleep(500)
        pin0.write_digital(0)
        sleep(500)
        pin0.write_digital(1)
        sleep(500)
        pin0.write_digital(0)
        sleep(500)
        pin0.write_digital(1)
        sleep(500)
        pin0.write_digital(0)
```

```
        sleep(1500)
        pin0.write_digital(1)
        sleep(1500)
        pin0.write_digital(0)
        sleep(500)
        pin0.write_digital(1)
        sleep(1500)
        pin0.write_digital(0)
        sleep(500)
        pin0.write_digital(1)
        sleep(1500)
        pin0.write_digital(0)
        sleep(1500)
        pin0.write_digital(1)
        sleep(500)
        pin0.write_digital(0)
        sleep(500)
        pin0.write_digital(1)
        sleep(500)
        pin0.write_digital(0)
        sleep(500)
        pin0.write_digital(1)
        sleep(500)
        pin0.write_digital(0)
        sleep(1500)
except KeyboardInterrupt as e:
    print("Interrupted by the user...")
```

We have now created the virtual SOS message.

Blinking two LEDs alternately

We can use pin 1 to connect another LED, as follows:

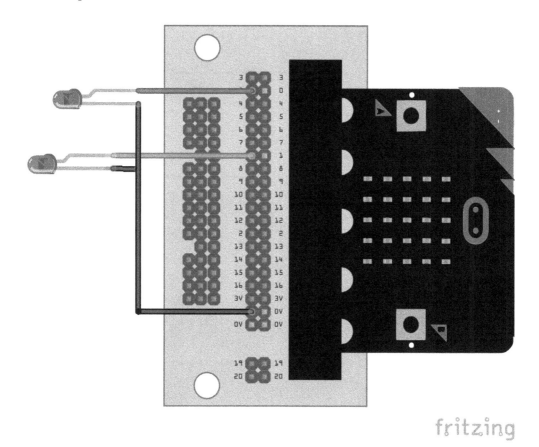

Figure 6.11 – Two LEDs connected to a Micro:bit

The following is the schematic of the circuit:

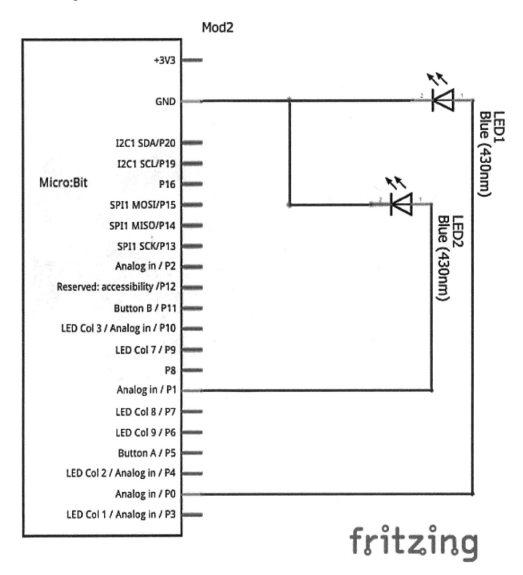

Figure 6.12 – Schematic for two LEDs connected to a Micro:bit

We can write a program to make the LEDs blink alternately, as follows:

```
from microbit import *
try:
    while True:
```

```
        pin0.write_digital(1)
        pin1.write_digital(0)
        sleep(500)

        pin1.write_digital(1)
        pin0.write_digital(0)
        sleep(500)

except KeyboardInterrupt as e:
    print("Interrupted by the user...")
```

If we run the code, we will see that the LEDs are blinking alternately. We can also write the following code to produce the same output as the preceding code but using the if condition:

```
from microbit import *
counter = 0
try:
    while True:
        if counter % 2 == 0:
            led1 = pin0
            led2 = pin1
        else:
            led1 = pin1
            led2 = pin0
        led1.write_digital(1)
        led2.write_digital(0)
        sleep(500)
        counter = counter + 1
except KeyboardInterrupt as e:
    print("Interrupted by the user...")
```

Again, this will make both LEDs blink alternately.

Traffic light simulator

We can also simulate traffic lights. For this, we will need three LEDs that are green, yellow (or orange), and red. Create a circuit, as shown in the following figure:

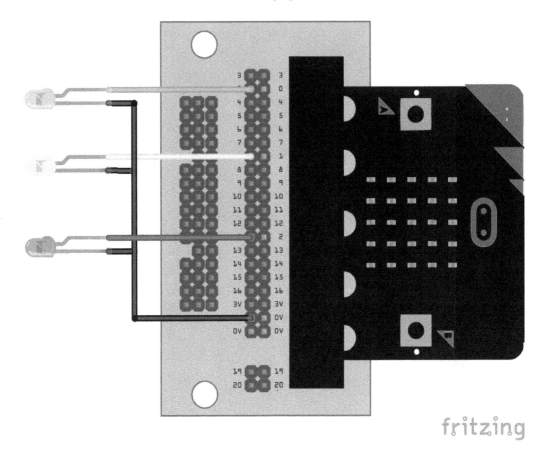

Figure 6.13 – Circuit diagram for a traffic light

The following is the schematic for the traffic light simulator:

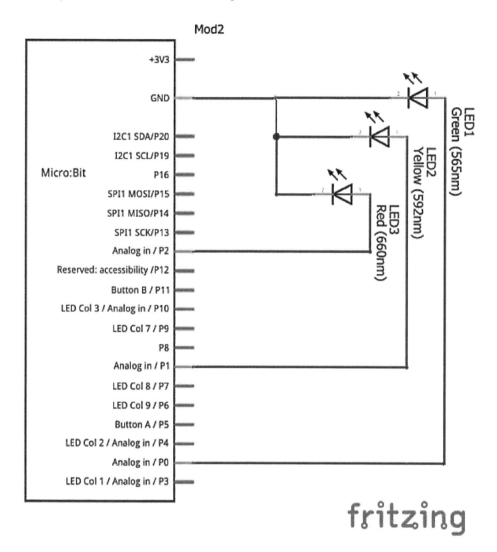

Figure 6.14 – Schematic for the traffic light simulator

The program is simple. I have created three user-defined functions that I am calling in an infinite loop, as follows:

```
from microbit import *
green = pin0
yellow = pin1
```

```
red = pin2
delay_green = delay_red = 5000
delay_yellow = 2000

def green_light():
    green.write_digital(1)
    yellow.write_digital(0)
    red.write_digital(0)

def yellow_light():
    green.write_digital(0)
    yellow.write_digital(1)
    red.write_digital(0)

def red_light():
    green.write_digital(0)
    yellow.write_digital(0)
    red.write_digital(1)

try:
    while True:
            green_light()
            sleep(delay_green)
            yellow_light()
            sleep(delay_yellow)
            red_light()
            sleep(delay_red)
except KeyboardInterrupt as e:
    print("Interrupted by the user...")
```

We have simulated traffic lights using the Micro:bit.

GPIO pins usage

Not all the GPIO pins can be programmed with MicroPython as a few of them are reserved. Let's demonstrate this. Run the following program:

```
from microbit import *
display.on()
try:
    while True:
        pin3.write_digital(1)
        sleep(500)
        pin3.write_digital(0)
        sleep(500)
except KeyboardInterrupt as e:
    print("Interrupted by the user...")
```

It will return the following error:

```
>>> %Run -c $EDITOR_CONTENT
Traceback (most recent call last):
  File "<stdin>", line 7, in <module>
ValueError: Pin 3 in display mode
```

This is because pin 3 is used by the built-in display matrix. This message about display mode is shown when we use pins 3, 4, 6, 7, and 10. This happens because these pins are connected to the built-in LED matrix of the Micro:bit. We will learn how to enable them at the end of the next section.

Pins 5 and 11 are used by two built-in push buttons labeled A and B. Trying to use these pins will show us the following error messages:

```
>>> %Run -c $EDITOR_CONTENT
Traceback (most recent call last):
  File "<stdin>", line 8, in <module>
ValueError: Pin 5 in button mode
>>> %Run -c $EDITOR_CONTENT
Traceback (most recent call last):
  File "<stdin>", line 9, in <module>
ValueError: Pin 11 in button mode
```

We cannot use these two pins (5 and 11) with LEDs using MicroPython.

Pins 17 and 18 are undefined as trying to use them in the program will result in the following errors:

```
>>> %Run -c $EDITOR_CONTENT
Traceback (most recent call last):
  File "<stdin>", line 9, in <module>
NameError: name 'pin17' isn't defined
>>> %Run -c $EDITOR_CONTENT
Traceback (most recent call last):
  File "<stdin>", line 9, in <module>
NameError: name 'pin18' isn't defined
```

Pins 19 and 20 are used for **Inter-Integrated Circuit** (**I2C**) buses and trying to use them in our program will result in the following errors:

```
>>> %Run -c $EDITOR_CONTENT
Traceback (most recent call last):
  File "<stdin>", line 8, in <module>
ValueError: Pin 19 in i2c mode
>>> %Run -c $EDITOR_CONTENT
Traceback (most recent call last):
  File "<stdin>", line 8, in <module>
ValueError: Pin 20 in i2c mode
```

Normally, we can use pins 0, 1, 2, 8, 9, 12, 13, 14, 15, and 16 directly for GPIO. We can also enable other pins, as we will see at the end of the next section.

4-bit binary counter

We can create a 4-bit binary counter with four LEDs (or a bigger counter with more LEDs). Create the circuit shown in the following diagram:

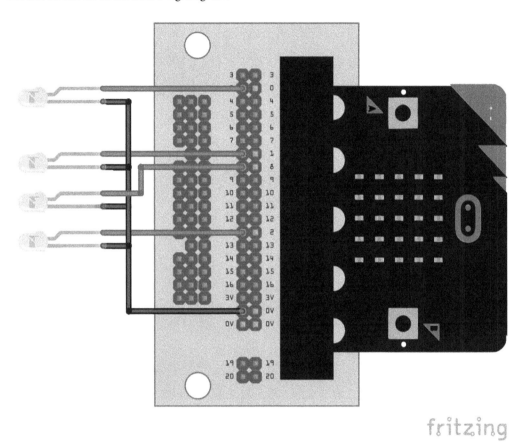

Figure 6.15 – Circuit for a 4-bit binary counter

Let's see the schematic for a 4-bit binary counter:

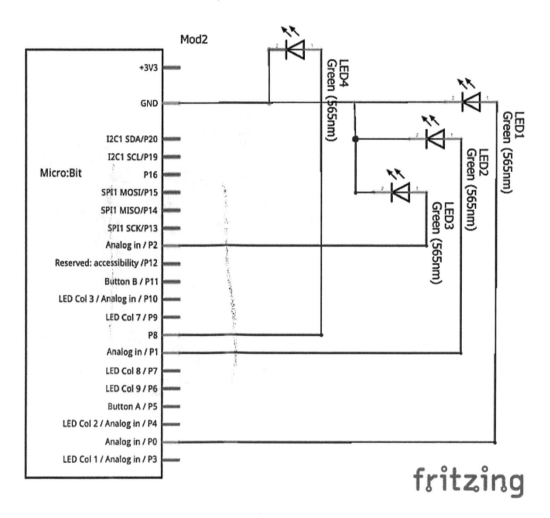

Figure 6.16 – Schematic for a 4-bit binary counter

Now, let's write the code for it:

```
from microbit import *
counter_led_pins = [pin0, pin1, pin2, pin8]
counter = 0
try:
    while True:
        for i in range(4):
```

```
            signal = int('{:04b}'.format(counter)[i])
            print(signal)
            counter_led_pins[i].write_digital(signal)
        print('------')
        if counter == 15:
            counter = 0
        else:
            counter = counter + 1
        sleep(1000)
except KeyboardInterrupt as e:
    print("Interrupted by the user...")
```

In this example, we created a list of the four pins. Then, we created a counter that cycles from 0 to 15. In each iteration of the loop, we compute the binary representation of the counter (using the `format()` method, where `{:04b}` means a 4-bit binary number representation) and assign every bit of that binary number to the group of pins. Thus, they will glow momentarily to represent the binary number. Pin 0 represents the most significant bit, while pin 8 represents the least significant bit of the binary representation of the counter. This is how we can write simple programs to demonstrate the working of LEDs.

Chaser effect

We can prepare a program for the chaser effect. The chaser effect is created by multiple LEDs blinking one after another in rapid succession. Connect LEDs to pins 0, 1, 2, 8, 9, 12, 13, 14, 15, and 16. I won't give you the circuit diagram this time as I believe that by now, you can comfortably understand how to prepare a simple circuit by following the description. We can blink the LEDs together with the following program:

```
from microbit import *

pins = [pin0, pin1, pin2,
        pin8, pin9, pin12,
        pin13, pin14, pin15,
        pin16]

try:
    while True:
        for pin in pins:
            pin.write_digital(1)
```

```
        sleep(1000)
        for pin in pins:
            pin.write_digital(0)
        sleep(1000)
except KeyboardInterrupt as e:
    print("Interrupted by the user...")
```

This program tests whether all the connections are correct. We can make simple modifications to it to create a simple chaser effect, as follows:

```
try:
    while True:
        for pin in pins:
            pin.write_digital(1)
            sleep(100)
        for pin in pins:
            pin.write_digital(0)
            sleep(100)
except KeyboardInterrupt as e:
    print("Interrupted by the user...")
```

Here, we have changed the positions of the call to the `sleep()` method, and we have also changed the arguments that are passed. The arguments that are passed to `sleep()` decide the speed of the effect.

We can also make all the LEDs blink one after another in rapid succession, as follows:

```
from microbit import *
pins = [pin0, pin1, pin2,
        pin8, pin9, pin12,
        pin13, pin14, pin15,
        pin16]


def blink(pin, duration):
    pin.write_digital(1)
    sleep(duration)
    pin.write_digital(0)
    sleep(duration)


try:
```

```
    while True:
        for pin in pins:
            blink(pin, 100)
except KeyboardInterrupt as e:
    print("Interrupted by the user...")
```

In this example, we created a custom function, blink(), and called it in a loop. We can make this effect bidirectional by modifying the loop in the following way:

```
try:
    while True:
        for pin in pins:
            blink(pin, 100)
        for pin in reversed(pins):
            blink(pin, 100)
except KeyboardInterrupt as e:
    print("Interrupted by the user...")
```

We can create a fancy effect by making the following modification to the loop:

```
try:
    while True:
        for i in range(0, len(pins), 1):
            blink(pins[i], 100)
            if not (i-1)<0:
                blink(pins[i-1], 100)
        for i in range(len(pins)-1, -1, -1):
            blink(pins[i], 100)
            if not (i+1)>8:
                blink(pins[i+1], 100)
except KeyboardInterrupt as e:
    print("Interrupted by the user...")
```

As we can see, the possibilities when writing code for this circuit are unlimited. I have used LEDs of a single color. However, you may wish to use LEDs of different colors to spice up the effect. You can also arrange them in a different geometric pattern.

Using an LED bar graph

Many manufacturers produce LEDs embedded in to a single package, known as **LED bar graph**, as shown in the following figure:

Figure 6.17 – An LED Bar Graph (courtesy: https://commons.wikimedia.
org/wiki/File:MFrey_LN3914N_AD-Converter.jpg)

In *Figure 6.17*, on the right-hand side, we can see an LED Bar Graph in action in an electronics project. It usually comes with 10 LEDs packaged in it. The LED Bar Graph has two rows of headers (like an integrated circuit) and fits perfectly around a breadboard's central groove. One row of the header is for applying the positive voltage, which should all be connected to GPIO pins. Another row must be connected to the ground. We have already created a circuit with 10 LEDs for the chaser effect. We can replace all 10 LEDs with a 10 LED Bar Graph. As an exercise, I built that circuit by replacing LEDs with the LED Bar Graph and ran all the earlier chaser programs with the new circuit to see the LED Bar Graph in action.

Enabling more pins

Earlier, we saw that pins 3, 4, 6, 7, and 10 could not be used directly in our programs for digital output as they are used by the built-in LED matrix. We can use them by disabling the matrix using the display.off() statement. So, before using these pins in our program, we just need to use that statement, and the pins can be used. The downside, however, is that we won't be able to use the built-in display in parallel. To turn the display on, we have to call `display.on()`. You may have noticed that we called this while using the display in *Chapter 5, Programming Built-In LEDs and Buttons*. It's always a good practice to explicitly call this before using the display, as the previous program may have turned it off.

In this section, we created the chaser effect with external LEDs and also learned how to enable the pins connected to the display by turning the display off.

RGB LEDs

RGB LEDs can emit three colors: red, green, and blue. They have four pins, one for each color; the remaining pin acts as a common cathode or common anode. The following circuit diagram shows the common cathode LED connection to the Micro:bit:

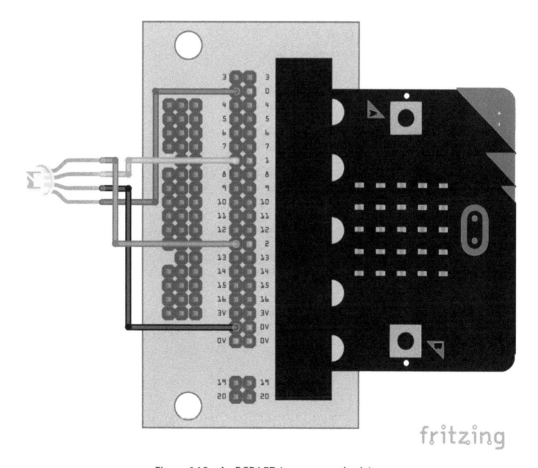

Figure 6.18 – An RGB LED (common cathode)

In a common cathode LED, we must apply a positive voltage to the color pins to activate the corresponding LEDs. We have connected the common cathode to the ground, the red pin to pin 0, the green pin to pin 1, and the blue pin to pin 2. The program that cycles through all the colors that this combination of digital outputs can produce is quite simple and intuitive, as follows:

```python
from microbit import *

red = pin0
green = pin1
blue = pin2

def glow(red_val, green_val, blue_val):
    red.write_digital(red_val)
    green.write_digital(green_val)
    blue.write_digital(blue_val)

try:
    while True:
        glow(0, 0, 0)
        sleep(1000)
        glow(0, 0, 1)
        sleep(1000)
        glow(0, 1, 0)
        sleep(1000)
        glow(0, 1, 1)
        sleep(1000)
        glow(1, 0, 0)
        sleep(1000)
        glow(1, 0, 1)
        sleep(1000)
        glow(1, 1, 0)
        sleep(1000)
        glow(1, 1, 1)
        sleep(1000)
except KeyboardInterrupt as e:
    print("Interrupted by the user...")
```

We can even modify the 4-bit counter program and turn it into a 3-bit counter (again, we are using `format()` and `{:03b}` to represent a 3-bit binary number). It will also cycle through all the possible combinations of colors. This is the program:

```
from microbit import *
rgb = [pin0, pin1, pin2]
counter = 7
try:
    while True:
        for i in range(len(rgb)):
            signal = int('{:03b}'.format(counter)[i])
            print(signal)
            rgb[i].write_digital(signal)
        print('------')
        if counter == 7:
            counter = 0
        else:
            counter = counter + 1
        sleep(1000)
except KeyboardInterrupt as e:
    print("Interrupted by the user...")
```

In a common anode RGB LED circuit, we need to connect the common anode pin to the 3V pin of the Micro:bit, as shown in the following circuit diagram:

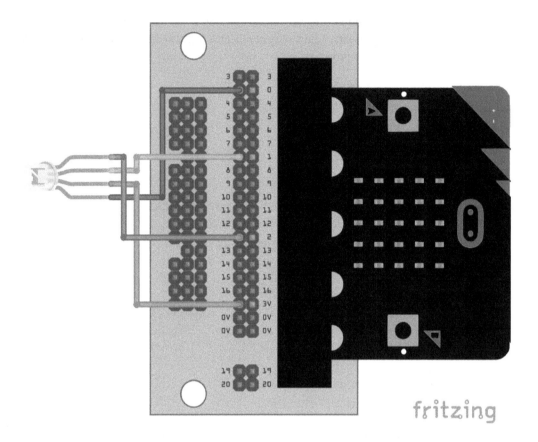

Figure 6.19 – An RGB LED (common anode)

The programs will remain the same. The preceding two programs will work perfectly for a common anode RGB LED. Only the order in which the colors are displayed will be reversed because, with common anode LEDs, we need to apply 0V to the color pins to activate the LEDs.

RGB LEDs are commonly used in maker projects. Knowing how to use them is advantageous. In this section, we learned about them in detail.

Seven-segment display

We can also work with special display devices that come with eight LEDs in them. They are known as seven-segment LEDs. They can display decimal numbers and decimal points. The following is a diagram of a common cathode seven-segment LED display:

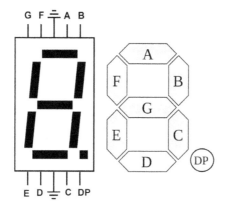

Figure 6.20 – A common cathode seven-segment LED display (courtesy: https://commons.wikimedia.org/wiki/File:7-Segment_Display_Visual_Pinout_Diagram.svg and https://commons.wikimedia.org/wiki/File:7_segment_display_labeled.svg)

We must apply a positive voltage to the corresponding pins and connect the common cathode pins to the ground. We can show numbers and a few alphabetical characters with a combination of pins. For example, if we activate pins A, B, C, D, E, and F together, it will show 0. Similarly, you can figure out what pins are to be activated for other numbers and characters. As an exercise for this chapter, connect this display to the Micro:bit and write a program that cycles through all the decimal and hexadecimal numbers in a continuous loop.

These displays are available in the common anode configuration too. The common anode pins must be connected to the 3V pin of the Micro:bit; to activate the LED, we have to send a LOW (that is, 0) signal to the corresponding LED. As an exercise, create a circuit for the common anode seven-segment LED display with the Micro:bit. You will need to make major modifications to the program you would write for a common cathode LED.

Summary

In this chapter, we started building circuits with external LEDs and derived electronic components. We have learned how to create interesting projects with our acquired knowledge.

In the next chapter, we will explore how to connect push buttons, slide switches, buzzers, and stepper motors to the Micro:bit and program them with MicroPython.

Further reading

You can find more information about GPIO pins and the Micro:bit-specific MicroPython API at https://microbit-micropython.readthedocs.io/en/v1.0.1/pin.html.

7

Programming External Push Buttons, Buzzers, and Stepper Motors

In the previous chapter, we worked with LEDs, seven segments, and their related components, observing them with the help of examples.

This chapter will explore using components such as switches, buzzers, and motors. We will explore their working principles and applications using the MicroPython programs and the Micro:bit. After completing this chapter, you will know how to use the following components:

- Push buttons
- Slide switches
- Buzzers
- Stepper motors

Let us get started by learning how to program these components.

Technical requirements

For this chapter, we will need the following hardware products:

- Push buttons
- LEDs
- A buzzer
- A stepper motor 28BYJ-48 and driver board (ULN2003A)

Push buttons

Push buttons are commonly used in switching operations. Electrical buttons are used in any electronics circuit when switching (on/off) functions need to be performed. Let us consider a scenario where an electrical bulb needs to be turned on or off; we usually go to a switchboard and press the button. It changes the state by moving to an open circuit. This can be done through the Micro:bit. The advantage it offers is that it can be controlled through the program.

The basic principle behind push buttons is to control the current's flow through pushing or pressing motions. A switch is considered on when it allows the flow of the current, so the circuit will be closed. In the off state, it breaks the current's flow, so the circuit will be open. There are various types of switches, as shown in *Figure 7.1*:

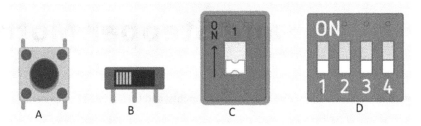

Figure 7.1 – Various types of switches – (A) push button, (B) slide
switch, (C) DIP DPST switch, and (D) DIP SPST switch x 4

The buttons mentioned in *Figure 7.1* can be used for various activities, as follows:

- Push button: To turn a single component attached to the system on and off.

- Slide switch: It has three terminals – the central pin is connected to the ground (GND), whereas the two side terminals are connected to two devices. It can control two components.

- **Dual In-Line Package (DIP) Double Pole Single Throw (DPST)** switch: The DIP specifies how a switch is grouped, as shown in *Figure 7.1 (C)* and *Figure 7.1 (D)*. DPST can separate power and ground and has an application in central power units. Here, poles refer to the number of connections in a switch. Throw refers to the position. Single throw means the position. Hence, DSPT allows you to throw two connections in the on and off state together.

- DIP **Single Pole Single Throw (SPST)**: DIP SPST is like DPST but with the difference being the number of devices in the input and output terminals. An SPST switch can only relate to one input and can only provide one output. As shown in *Figure 7.1 (D)*, the four SPST switches are formed in DIPs, which is why DIP SPST switch x 4 has been specified. In xN, N represents the number of switches available in a DIP.

Now that we've looked at the different types of switches, let us explore how to use them with the Micro:bit. We explored built-in push buttons in detail in *Chapter 5, Built-in LED Matrix Display and Push Buttons*. We'll revise their usage in this section. As shown in *Figure 7.2*, the two circles represent the push buttons that come built-in with the Micro:bit. These buttons are named A (on the left-hand side) and B (on the right-hand side) on the board. For programming purposes, we will use A and B as the names of the buttons:

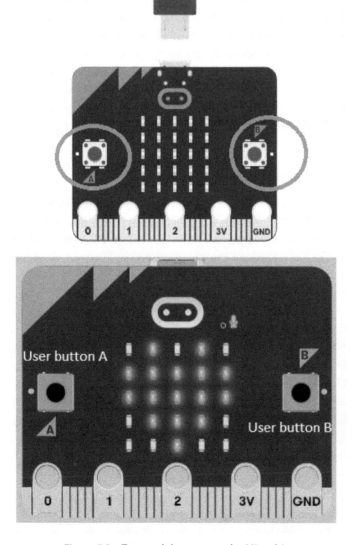

Figure 7.2 – Two push buttons on the Micro:bit

To program these buttons, we need to recognize the push operation. The following program has three main steps, which include initialization, loop formation, and providing the logic:

```
from microbit import *
display.clear()
while True:
    if button_a.is_pressed(): #returns true or false
        display.show(Image.GHOST)
    elif button_b.is_pressed():#returns true or false
            display.show(Image.SKULL)
```

In the preceding program, after importing the libraries, display.clear() will clear any content on the LED panel. In the while loop, the conditions of buttons A and B are mentioned. display.show() will show a ghost image when button A is pressed, and a skull image will appear when button B is pressed.

Connecting an external push button

To understand the usefulness of the external push button, let us look at an example. We would like to connect and control an LED's on/off state using an external push button. To do so, we need a push button and an LED connected to the Micro:bit, as shown in *Figure 7.3*. Here, we can see that a push button is a four-legged device representing two pins as GND and two pins for connecting a device:

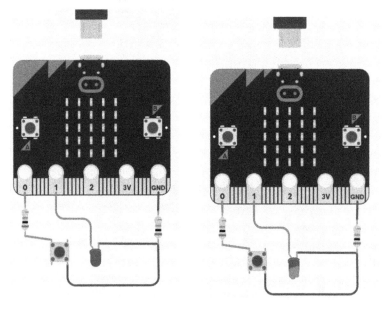

Figure 7.3 – Connecting a push button and its output via a push operation

We don't need to differentiate between the four pins in a push button since they are provided to a static connection. This is because they are most commonly used in Printed Circuit Boards (PCBs) for reset and power-on operations. One push button can control many devices. Once connected to GND and pin 0, the push button connection is made. Now, we need to connect an LED to another pin of the Micro:bit. As shown in *Figure 7.3*, it is connected to pin 1. The following program will enable the read pin and write pins based on the status of the button state – that is, on or off:

```
from microbit import *

while True:
    if pin0.read_digital():
        pin1.write_digital(1)
        sleep(1000)
    else:
        pin1.write_digital(0)
        sleep(500)
```

To implement this functionality, we need to write a program. In this example program, the `if-else` condition has been provided, which makes the circuit work on both `True` and `False` states. `pin0.read_digital():` is a pin connected to the button, which indicates that if the switch is pressed, then the LED connected to pin 1 should be on. If the push button is not pressed, `pin1.write_digital(0)` will turn the LED off. With the help of code, a user can connect the push button and the LED to various pins of the Micro:bit board by simply replacing the pin's number.

In this section, we learned how to use push buttons using the Micro:bit. We observed the working principles, classifications, and applications of switches. We also learned how to interface the push button with the help of a circuit diagram and program. In the next section, we will learn how to use slide switches.

Slide switches

A slide switch has three terminals and can be used to control two components. It is a type of SPST with two states where external devices can be connected. Its applications include two-way switching, home appliance control, office power control, and others. The functioning of the slide switch can be understood with the help of *Figure 7.4*:

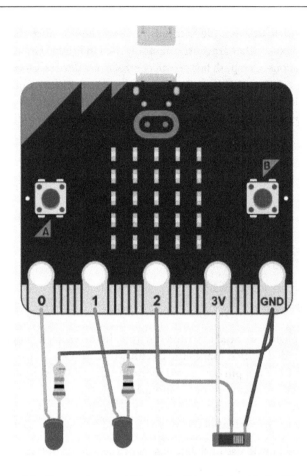

Figure 7.4 – Connecting two LEDs and a slide switch with the Micro:bit

In *Figure 7.4*, a slide switch is connected to pin 2 with GND and VCC connections. To observe its working, we have connected two LEDs to pins 0 and 1. The important thing to observe here is that two resistors are also connected to the LEDs. These resistors help control the voltage level of LEDs so that their voltage and current levels don't vary.

In the following program, we are controlling two LEDs with one switch. Here, we need to assign the LED connections and the switch connection. The rest will be based on the position of the slide switch; it will turn on one LED at a time:

```
from microbit import *
while True:
    if pin2.read_digital():
        pin0.write_digital(1)
```

```
        pin1.write_digital(0)
        sleep(1000)
    else:
        pin1.write_digital(1)
        pin0.write_digital(0)
        sleep(1000)
```

The preceding program has `pin1.write_digital()` and `pin0.write_digital()` to control the on and off states of both LEDs, where 1 is for on and 0 is for off, respectively. The program works almost the same as it did in the case of one switch and one LED, except that in this case, two LEDs are connected, and the same is reflected in the program.

Counting how many times a button is pushed

The good thing about a programable switch is that we can not only control the on-off state but also operate based on the number of times a button is pressed. In the following program, a push button is being used and a counter has been set to count the number of times a push button switch is pressed:

```
from microbit import *
counter = 0
display.show(counter)
while True:
    if pin0.read_digital():
        counter = counter + 1
        display.show(counter)
        pin1.write_digital(1)
        sleep(1000)
    else:
        pin1.write_digital(0)
        sleep(1000)
```

In the preceding program, we can see that the counter variable has been assigned an initial value of 0. In the while loop, a switch connected to pin 0, as shown in *Figure 7.4* is connected, and its state is being monitored through `if-else` statements. The counter shows the count, which is initially set to zero. On every press, the counting will start with an increment of +1:

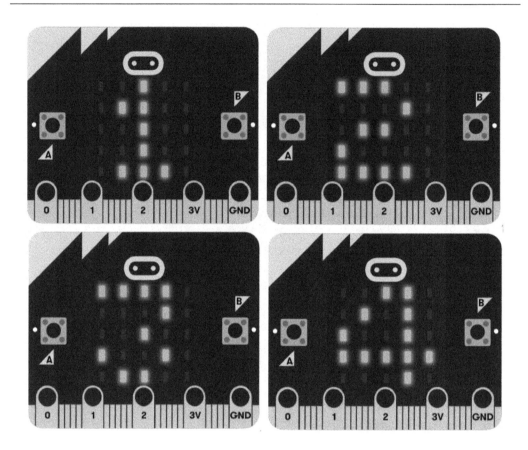

Figure 7.5 – Illustrating the results of counting the number of times a button is pressed

This can be observed in *Figure 7.5*, which shows that after every press, the LED display is incremented by one. On the coding side, we need to assign a counter and some logic. The counter will help in measuring the number of times the button is pressed. The counter will be part of the logic, as shown in the form of an if-else condition:

```
from microbit import *
counter = 10
display.show(counter)
while True:
    if pin0.read_digital():
        counter = counter - 1
        display.show(counter)
        pin1.write_digital(1)
        sleep(1000)
```

```
else:
    pin1.write_digital(0)
    sleep(1000)
```

The preceding code shows the down count after going through the same hardware configuration on the hardware side and setting the counter to an initial value of 10. In code, we can do the down count by using counter = counter - 1.

Connecting multiple push buttons

If we want to try and make a slightly complex design using multiple push buttons and control the LEDs, we need to make specific changes to the hardware and program:

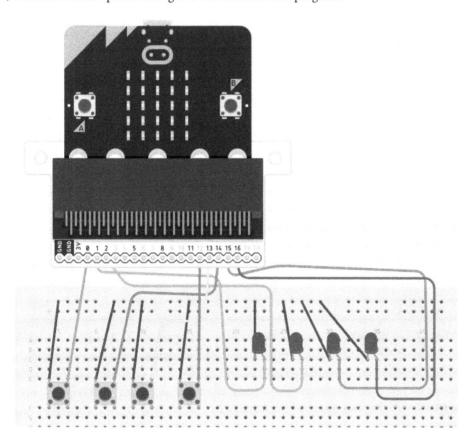

Figure 7.6 – Controlling multiple switches and LEDs connected to the Micro:bit

Figure 7.6 shows the switches are connected to pins 0, 12, 13, and 14, whereas the LEDs are connected to pins 1, 2, 15, and 16. Keep in mind that all the devices have a common ground connection.

The program logic will be more or less the same as controlling one switch and LED, except that in this case, we have multiple input-output devices. As per our understanding, separate pins are assigned to switches and LEDs. This is where the on and off states will be defined:

```
from microbit import *
while True:
    if pin0.read_digital():
        pin1.write_digital(1)
        pin2.write_digital(0)
        pin15.write_digital(0)
        pin16.write_digital(0)
        sleep(1000)
    elif pin14.read_digital():
        pin1.write_digital(0)
        pin2.write_digital(1)
        pin15.write_digital(0)
        pin16.write_digital(0)
    elif pin12.read_digital():
        pin1.write_digital(0)
        pin2.write_digital(0)
        pin15.write_digital(1)
        pin16.write_digital(0)
    elif pin13.read_digital():
        pin1.write_digital(0)
        pin2.write_digital(0)
        pin15.write_digital(0)
        pin16.write_digital(1)
```

In the preceding program, pin0.read_digital():, pin14.read_digital():, pin13.read_digital():, and pin13.read_digital(): are connected to the switch and pin1.write_digital(0), pin2.write_digital(0), pin3.write_digital(0), and pin4.write_digital(0) are write options for the LEDs. In pinN.write_digital( ), the 0 and 1 indicate the on and off states, respectively.

The if-elif conditions will execute the respective part of the code when the condition is true. In this way, we are turning on one LED per switch. For example, in the if pin0.read_digital(): condition, write_digital is set as pin1.write_digital(1), pin2.write_digital(0), pin3.write_digital(0), and pin4.write_digital(0), which indicates that only pin 1 is on and the rest are off for the LEDs.

In this section, we covered the slide switch, counted the number of times a switch is pressed, and observed how to handle multiple input-output devices using the Micro:bit. In the next section, we will cover buzzers and their applications.

Buzzers

A buzzer is a piezoelectric device that generates sound. Buzzers have multiple applications, such as generating an audio signal or acting as alarms and beepers. *Figure 7.7* shows a buzzer device with positive and negative signal pins:

Figure 7.7 – A buzzer

In the Micro:bit V2, a speaker is connected to the board, as shown in *Figure 7.8*. It is connected on the flip side of the LED array. From *Figure 7.8*, we can see that the speaker/buzzer is connected close to the processor:

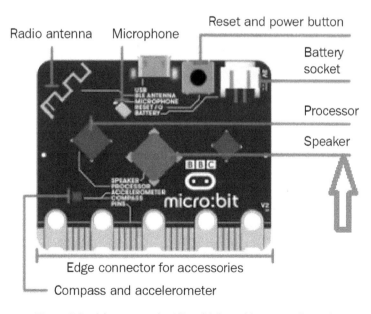

Figure 7.8 – A buzzer on the Micro:bit board (courtesy: https://
microbit.org/get-started/user-guide/overview/)

The onboard buzzer can be used directly by executing the following simple program:

```
import music
music.play(music.ODE)
```

In the preceding code, the `import music` module contains predefined tunes that can be used to generate sound. `music.play(music.ODE)` can play the music through the buzzer:

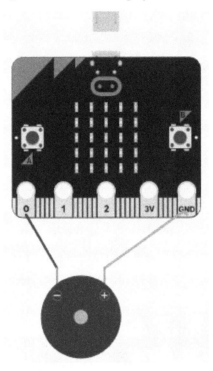

Figure 7.9 – Connecting an external buzzer to the Micro:bit

The user can also connect an external buzzer to the Micro:bit board. The default pin that's assigned to connect a buzzer or a speaker is pin 0. As shown in *Figure 7.9*, the positive pin of the buzzer is connected to GND, while the negative pin is connected to pin 0. Now, let us set the buzzer to play some tunes by simply mentioning them in the following program:

```
from Micro:bit import *
import music
tune = ["A4:4", "B4:4", "C4:4", "D4:4", "C4:4", "D4:4",
"E4:4",  "C4:4", "E4:4", "F4:4", "G4:8", "E4:4", "F4:4",
"G4:8"]
music.play(tune)
```

In the preceding program, a few chords have been mentioned in the tune variable. The chords are evoked through import music, and all the chords mentioned in the tune variable will be played in sequence repetitively.

Let us try to connect a buzzer and switches with the Micro:bit board. To do so, we need three signal pins – that is, pin 0 will be connected to the buzzer, pin 1 will be connected to switch 1, and pin 2 will be connected to switch 2. The ground terminal of all three devices will be connected to the GND pin of the Micro:bit board, as shown in *Figure 7.10*:

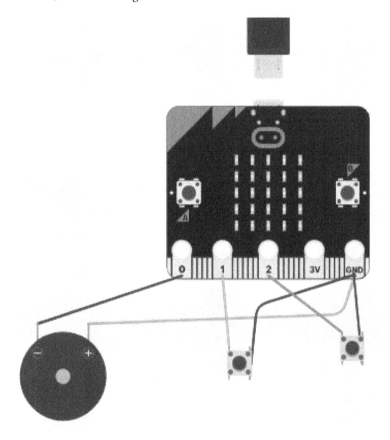

Figure 7.10 – Connecting the switches and a buzzer

Now, we will try to play some audio speech and tunes using a buzzer with the help of our program. First, we need to import the speech and music libraries:

```
from microbit import *
import music
import speech
while True:
    if pin2.read_digital() == 1:
        music.play("A2:5")
        speech.say("Hello Abhishek")
    elif pin1.read_digital() == 1:
        music.play("E1:7")
        speech.say("MicrooooooooooooooooooPython")
    else:
        pass
```

In the preceding program, we can see that two libraries are being used: import music and import speech. In a while loop, pins 1 and 2 are assigned as digital read using read_digital. In the if-elif conditions, switch 1 and switch 2 are selected, and for both selections, different music tones and speech are assigned. The speech functions are evoked through speech.say().

In this section, we looked at a few applications of buzzers using the Micro:bit and played musical tunes to generate speech. We also learned how to use the music and speech libraries. In the next section, we will be using a stepper motor and looking into its applications.

Stepper motors

A stepper motor is the most commonly used motor in line manufacturing production units and industrial applications. It provides precision with angular motion and torque. We will use the 28BYJ-48 stepper motor for this example. This type of motor includes two main components. The stators are magnetic coils that create an electromagnetic field, and the rotor is a metal shaft that rotates based on the electric field generated:

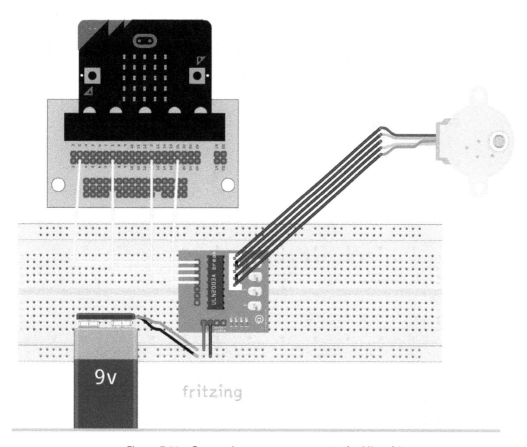

Figure 7.11 – Connecting a stepper motor to the Micro:bit

As shown in *Figure 7.11*, a Micro:bit board is connected to the stepper motor through a driver. A 9V battery is connected to the driver circuit to produce driving power, and the five pins of the stepper motor are connected to the driver board. As we can see, the stepper motor is connected to the Micro:bit through the driver board. It is important to understand the core functionality of the driver circuit. ULN2003A has a pair of Darlington transistors that help with load management as the stepper motor consumes heavy current values. IN pins are control inputs from the driver board that go through the Micro:bit pins. In *Figure 7.11*, we are using pins 0, 1, 2, and 14 to control the stepper motor. The programming of the stepper motor includes pin initialization, fixing the start and stop values, and turning on the stators one at a time:

```
from microbit import *
import random
pins = [pin0, pin1, pin2, pin16]
while True :
```

```
cycle = random.randrange(Start, Stop)
while True:
    pins[0].write_digital(1)
    pins[1].write_digital(0)
    pins[2].write_digital(0)
    pins[3].write_digital(0)
    sleep(10)
    pins[0].write_digital(0)
    pins[1].write_digital(1)
    pins[2].write_digital(0)
    pins[3].write_digital(0)
    sleep(10)
    pins[0].write_digital(0)
    pins[1].write_digital(0)
    pins[2].write_digital(1)
    pins[3].write_digital(0)
    sleep(10)
    pins[0].write_digital(0)
    pins[1].write_digital(0)
    pins[2].write_digital(0)
    pins[3].write_digital(1)
    sleep(10)
```

In the preceding program, the rotation has been set to import random and will generate a random number, which will be used to perform a rotatory operation. The four pins connected to the Micro:bit are enabled in one-at-a-time mode at a given time step and will be controlled.

In this section, we learned about hardware connections and programming a stepper motor to control its rotatory motions. By changing the start and stop values, users can observe the change in the motions. Apart from that, users can also control the speed and angle of the stepper motor.

Summary

In this chapter, we explored the different types of external buttons that are available and their functions. First, we learned how to combine switches with devices such as LEDs. Then, we learned how to combine a buzzer with a switch. We also explored the motor driver circuit and the stepper motor. Finally, we explored commonly used libraries such as music, speech, and random.

In the next chapter, we will investigate the filesystem of the Micro:bit using MicroPython and its applications, such as file handling methods, OS modules, and MicroFS modules.

Further reading

You can find out more about buzzers, switches, and motor drivers for the Micro:bit's implementation of MicroPython at https://microbit-micropython.readthedocs.io/en/latest/tutorials/buttons.html, `https://microbit-micropython.readthedocs.io/en/latest/speech.html`, and `https://microbit-micropython.readthedocs.io/en/latest/music.html`, respectively.

Part 3: Filesystems and Programming Analog I/O

In this section, you will discover how to explore the filesystem. You will explore the applications of analog input and **Pulse Width Modulation (PWM)**. You will understand how to accept analog input from potentiometers and photoresistors. You will also employ PWM to work with servo motors and RGB LEDs.

This section has the following chapters:

- *Chapter 8, Exploring the Filesystem*
- *Chapter 9, Working with Analog Input and PWM*

8

Exploring the Filesystem

In the previous chapter, we worked with push buttons, slide switches, buzzers, and stepper motors. We used them with the Micro:bit and MicroPython. This chapter is a little bit of a detour from the hardware components we have been exploring.

This chapter focuses on the filesystem of Micro:bit. We will fully explore all the functionality and use cases pertaining to the filesystem. Here is a list of the topics we will cover in this chapter:

- Creating and reading files
- Appending a file
- Creating our own library
- The OS module
- Working with MicroFS

Let's explore working with files with MicroPython and Micro:bit.

Technical requirements

We just need the usual setup of a Micro:bit connected to a computer with internet connectivity.

Creating and reading files

Micro:bit has a persistent filesystem. This means that the files remain intact even when we power it off. Micro:bit stores all the files in the flash memory. We can use approximately 30 KB of flash memory to store our own files. One of the constraints of the Micro:bit filesystem is that it provides a **flat** filesystem. This means that we cannot create directories (folders) and subdirectories. It won't allow us to create a hierarchy of directories and files as commonly found in operating systems such as Windows, Unix, FreeBSD, and Linux. The stored files will remain intact until we delete them either manually or programmatically. Also, re-flashing the Micro:bit deletes a file(s).

We can create and store files with any extension on Micro:bit. We can use the built-in open() function to create a file in various modes. We have to specify whether we wish to read or write the file in the arguments. The 'w' argument stands for **write mode**. Here is a simple example:

```
with open('HenryV.txt', 'w') as file_handle:
    file_handle.write("We few, we happy few, we band of
brothers;")
    file_handle.write("\nFor he to-day that sheds his blood
with me")
    file_handle.write("\nShall be my brother; be he ne'er so
vile,")
    file_handle.write("\nThis day shall gentle his condition;")
    file_handle.write("\nAnd gentlemen in England now a-bed")
    file_handle.write("\nShall think themselves accurs'd they
were not here,")
    file_handle.write("\nAnd hold their manhoods cheap whiles
any speaks")
    file_handle.write("\nThat fought with us upon Saint
Crispin's day.")
```

Run the code in Thonny and go to the main menu of the IDE. Click on **View** and the **Files** option under that menu. Check the following screenshot for guidance:

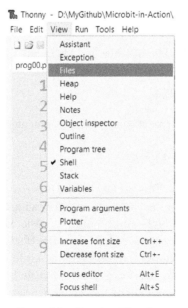

Figure 8.1 – Opening Files view

It opens a view, as shown in the following screenshot:

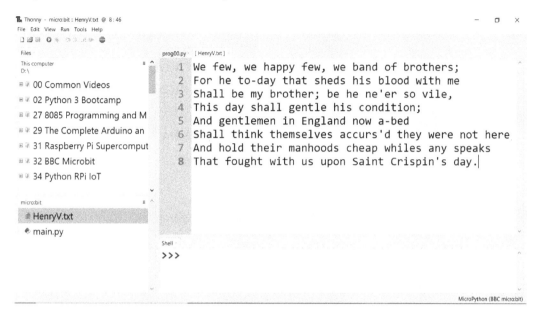

Figure 8.2 – The Files view

In *Figure 8.2*, we can see the filesystem of our computer in the top-left panel. We can browse files here. Following that is the filesystem of the Micro:bit. We can see the main.py file. When we flash any Python file to Micro:bit using the Mu editor, it saves the code in this file. We have to save our code with this filename manually while using the Thonny editor. If we wish the Micro:bit to run a program at power-up or reboot, then we need to either flash the program using the Mu editor (which will automatically write the program to the main.py file on the Micro:bit) or use the Thonny IDE to save the program manually on the Micro:bit with the filename as main.py. I have opened the HenryV.txt file by double-clicking it. We can see the text in the editor. We can even delete the files on the Micro:bit by right-clicking the file and choosing the appropriate options. Explore this interface yourself further as an exercise.

The constraint with the **Files** view is that it does not work when a program is under execution. The files on the Micro:bit are not visible when a program is running, as shown in the following screenshot:

Figure 8.3 – The Files view when a program is executing in Thonny

Stopping the program under execution will show the files currently on the Micro:bit.

The Mu editor also has a **Files** view. We can invoke the view using the **Files** button in the menu. The following screenshot shows the **Files** view:

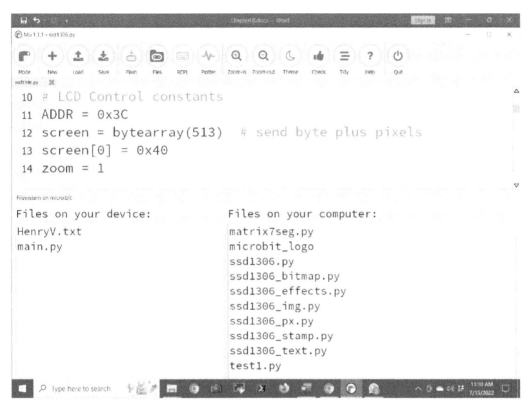

Figure 8.4 – The Files view when a program is executing in the Mu editor

The panels in the lower half show the files on the device (Micro:bit, in this case) and on your computer. However, there is a catch. Unlike the Thonny IDE, the Mu editor **Files** view cannot browse the files on your computer. The Mu editor creates a directory in the home directory of the user of your operating system. The directory is named mu_code. Currently, I am using MS Windows, and the directory for the Mu editor is located at C:\Users\Ashwin\mu_code. The Mu editor shows only the files in this directory. We can drag and drop files to copy files between our computer and the Micro:bit. We can even delete files on the Micro:bit.

I prefer to use the Thonny IDE. However, we can use any editor for exploring files and programming the Micro:bit. I like the Thonny IDE as we can run a program on the Micro:bit's main memory directly from the IDE without flashing it to the main.py file on the Micro:bit. The Mu editor does not have this facility.

Let's programmatically read the data stored in this file. Here is the code:

```
with open('HenryV.txt', 'r') as file_handle:
    print(file_handle.read())
```

The open () function opens any file by default in read mode, so the 'r' argument is optional. We can write the preceding example, as follows:

```
with open('HenryV.txt') as file_handle:
    print(file_handle.read())
```

Both the preceding examples read the contents of the file passed as an argument and show it in the REPL.

This is how we can create and read files with MicroPython and BBC Micro:bit.

Appending a file

The standard Python implementation (CPython) has the provision for appending a file. If we open a file again in write mode and add some information, earlier information is erased, and new information is overwritten. However, MicroPython does not have this functionality. We can, however, append a file with a clever trick. We first open the file to be appended in read mode and save the contents to the variable. Then, we add another string to that variable and write the variable to the file to be appended. This will append the file. The following program demonstrates that:

```
with open('HenryV.txt') as file_handle:
    file_data = file_handle.read()
file_data = file_data + '\nHenry V, William Shakespeare'
with open('HenryV.txt', 'w') as file_handle:
    file_handle.write(file_data)
with open('HenryV.txt') as file_handle:
    print(file_handle.read())
```

Creating our own library

We can create our own library. Its functionality can be used in other Python programs and in the REPL tool. Let's try that. Save the following program with the name mylib.py on your local computer:

```
def message():
    print("Then out spake brave Horatius,")
    print("The Captain of the Gate:")
    print("\"To every man upon this earth")
    print("Death cometh soon or late.")
    print("And how can man die better")
    print("Than facing fearful odds,")
    print("For the ashes of his fathers,")
    print("And the temples of his Gods.\"")
```

```
if __name__ == '__main__':
    message()
```

The `if` condition here is to check if we care about running the program directly or calling it with an external module. If we run the program directly, it runs the code enclosed under that. This is useful for writing the test code for our program. A Python program is also known as a module. Run the preceding program/module and check for any errors. Once we verify the program, we have to save it on the Micro:bit with the same name, `mylib.py`. Now, write the following statements in the REPL:

```
>>> import mylib
>>> message()
```

It will print all the output of the `print()` statements in the REPL. You can also save these statements as a Python file.

This is the simplest way we can create and use our own library.

The OS module

We can perform many operations on files and get a lot of information about the system using the OS module. Let's use the REPL to check the functionality offered by the OS module. Let's import it:

```
>>> import os
```

We can retrieve the properties, as follows:

```
>>> os.uname()
(sysname='microbit', nodename='microbit', release='2.0.0',
version='Micro:bit v2.0.0+b51a405 on 2021-06-30; MicroPython
v1.15-64-g1e2f0d280 on 2021-06-30', machine='Micro:bit with
nRF52833')
```

We can retrieve a list of current files on the Micro:bit's flash memory:

```
>>> os.listdir()
['HenryV.txt', 'mylib.py', 'main.py']
```

We can remove a file, as follows:

```
>>> os.remove('HenryV.txt')
>>> os.listdir()
['mylib.py', 'main.py']
```

We can see the size, as follows:

```
>>> os.size('mylib.py')
399
```

We can check the statistics of a file:

```
>>> os.stat('mylib.py')
(32768, 0, 0, 0, 0, 0, 399, 0, 0, 0)
```

Readers are encouraged to check the Micro:bit MicroPython documentation for all the functions used in these examples for more details. This is how we can work with the built-in OS module in Micro:bit. Let's see how to work with files using **MicroFS** in the next section.

Working with MicroFS

MicroFS is a simple command-line tool to interact from your computer with the filesystem of Micro:bit. It works with the command line interpreter program of your computer's operating system (MS Windows, Linux, FreeBSD, and macOS). We must download it first from https://www.python.org/downloads/ and then install it. While installing, check all the options, as shown in the following screenshot:

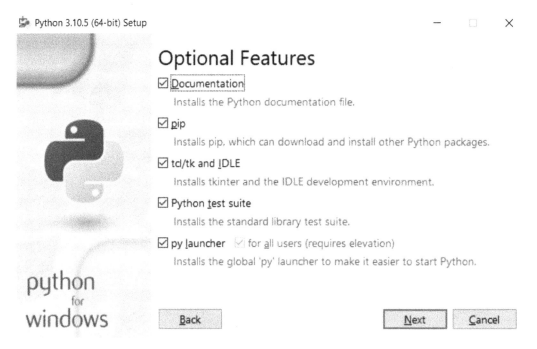

Figure 8.5 – Installing Python, pip, and IDLE

You also need to check all the options on the next screen of the installation wizard, as shown in the following screenshot:

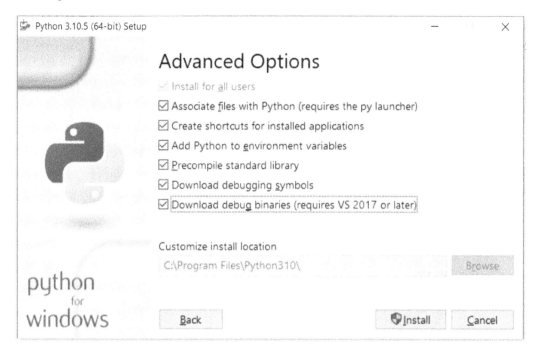

Figure 8.6 – Advanced Options

This will install Python, **pip** (**pip installs Python**; it's a recursive acronym, a technical pun), and **Integrated Development and Learning Environment** (**IDLE**). pip is a **package manager** for Python. It will also add Python and pip to the list of environment variables, enabling us to launch it directly from any command line tool (such as CMD or PowerShell). We can run the following command on the command line interpreter (*NOT* the Python shell or the MicroPython REPL) of your computer's operating system to install MicroFS with pip:

```
pip3 install microfs
```

This will install the MicroFS command-line utility and Python library to your computer (*NOT* Micro:bit). Let's see a demonstration of the utility. Open the command line interpreter on your computer as an administrator (I am using MS Windows PowerShell as an administrator) and run the following command to see a list of files on the Micro:bit connected to your computers:

```
PS C:\Users\Ashwin> ufs ls
```

It shows the current files, as follows:

```
mylib.py main.py
```

We can copy a file from the Micro:bit to the current directory of the command line interpreter, as follows:

```
PS C:\Users\Ashwin> ufs get mylib.py
```

We can now see the file in the current directory of the command line interpreter as follows (using the ls command if your command line interpreter program supports it—for example, PowerShell and Bash terminal):

```
PS C:\Users\Ashwin> ls mylib.py
    Directory: C:\Users\Ashwin
Mode                 LastWriteTime         Length Name
----                 -------------         ------ ----
-a---         7/15/2022   4:29 PM            399 mylib.py
```

We can also copy a local file to the Micro:bit, as follows:

```
PS C:\Users\Ashwin> ufs put test.txt
```

We can use the following command again to see the files of the Micro:bit:

```
PS C:\Users\Ashwin> ufs ls
mylib.py main.py test.txt
```

We can remove a specific file from the Micro:bit, as follows:

```
PS C:\Users\Ashwin> ufs rm test.txt
PS C:\Users\Ashwin> ufs ls
mylib.py main.py
```

We can also write a Python program using the installed MicroFS module and execute it using IDLE on the local computer (*NOT* MicroPython on the Micro:bit). Here is the code for the program:

```
import microfs as ufs


print(ufs.ls())


try:
    ufs.get('mylib.py')
```

```
    print("File is copied to the local computer successfully!")
except Exception:
    print("No such file exists on the connected Micro:bit
device!")

try:
    ufs.put('prog06.py')
    print("File is copied to the Micro:bit successfully!")
except Exception:
    print("No such file exists on the local computer!")

print(ufs.ls())

try:
    ufs.rm('prog06.py')
    print("File is deleted from the Micro:bit successfully!")
except Exception:
    print("No such file exists on the connected Micro:bit
device!")

print(ufs.ls())
```

Run the program on IDLE on your local computer while the Micro:bit is connected to it, and it shows the following output:

```
['mylib.py', 'main.py']
File is copied to the local computer successfully!
File is copied to the Micro:bit successfully!
['mylib.py', 'main.py', 'prog06.py']
File is deleted from the Micro:bit successfully!
['mylib.py', 'main.py']
```

This is how we can work with the filesystem of Micro:bit with the MicroFS module installed on our local computer.

Summary

In this chapter, we learned how to work with the limited filesystem of Micro:bit powered by MicroPython. We learned how to create, delete, read, and alter the contents of files on Micro:bit using MicroPython programs. We also learned how to work with the OS module in MicroPython and the MicroFS module of Python on our computer. We will use this knowledge in the upcoming chapters of the book when we need to install and work with external libraries for external hardware modules.

In the next chapter, we will explore the concepts of analog input and **pulse-width modulation** (**PWM**). Using these concepts, we will program a potentiometer, photoresistor, servo motor, **red, blue, and green light-emitting diode** (**RGB LED**), and joystick with microPython on Micro:bit.

Further reading

- You can find more information about filesystems and MicroPython programming at `https://microbit-micropython.readthedocs.io/en/v1.0.1/tutorials/storage.html`.

- You can find more information about the Micro:bit-specific OS module of MicroPython at `https://microbit-micropython.readthedocs.io/en/v1.0.1/os.html`.

- You can explore more information about MicroFS at the GitHub page of the project at `https://github.com/ntoll/microfs`.

9

Working with Analog Input and PWM

In the previous chapter, we learned how to work with the files stored on Micro:bit using MicroPython.

This chapter will explore how to use analog components and interface them to analog pins on the Micro:bit. We will also explore using various analog devices, such as a potentiometer, a joystick, a **Pulse-Width Modulation** (**PWM**) signal, a servo, a photoresistor, and various functions supported by Micro:bit for analog applications.

In this chapter, we will cover the following topics:

- Micro:bit analog pins
- Potentiometers
- Photoresistors
- PWM signals
- Servo motors using PWM
- Handling multiple analog devices
- PWM using RGB LED
- Joysticks

Let's explore the analog input and PWM with Micro:bit and MicroPython.

Technical requirements

For this chapter, we will need the following hardware products:

- A potentiometer

- A servo motor

- A photoresistor

- A joystick

Micro:bit analog pins

Besides having onboard sensors, micro: bit also facilitates integrating sensors using pins. The sensors can be connected as an input device (i.e., to feed the data into the Micro:bit – for example, a temperature sensor) or as an output device (i.e., the data or command coming from the Micro:bit to affect the outside world, such as operating a motor). The analog signal plays a vital role in handling the sensors. In *Figure 9.1*, the Micro:bit board is presented, showing the analog pins – that is, P0, P1, P2, P4, and P10. These pins are used to handle analog inputs:

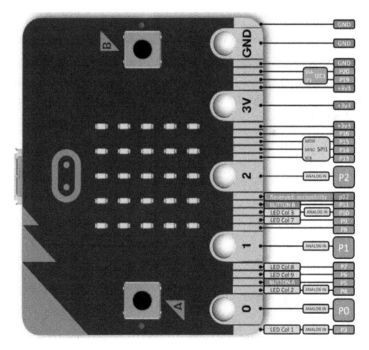

Figure 9.1 – Analog pins in Micro:bit (courtesy: https://microbit-micropython.readthedocs.io/en/v1.0.1/pin.html)

The analog signals are continuous signals represented in the form of sinusoidal waves with a continuous range of values. By contrast, digital signals are represented by square waves and have values of 1 or 0 (high or low). The pin selection is based on understanding the type of sensors used for an application. If the sensor is analog, then it needs to be connected through analog pins, and if it is digital, then it relates to digital pins, as mentioned in *Figure 9.1*.

In this chapter, we will look into handling analog devices using MicroPython. The Micro:bit supports the following analog operations:

- `pin.read_analog()`: Reads the value coming from the sensor connected to a respective pin. For example, `pin0.read_analog()` represents reading the value of a sensor connected to pin number 0 – that is, P0 as mentioned in *Figure 9.1*

- `pin.write_analog()`: Puts the continuous value to the pin so that the sensor connected to that pin can receive it. For example, `pin14.write_analog()` means that the values are coming as an output to pin number 14 or P14.

- `pin.is_touched()`: If the selected pin is touched, it returns `True`; if not, it returns `False`. For example, `pin0.is_touched()` signifies that if P0 is touched or any resistance is created, it returns `True`.

- `set_analog_period(period)`: Used for PWM to set a signal as an output in milliseconds. For example, `pin1.set_analog_period(2)` indicates that pin 1 is the PWM signal output for 2 milliseconds. Note that the minimum value of `(period)` should be 1 millisecond.

- `set_analog_period_microseconds(period)`: Has same functionality as `set_analog_period(period)` except that the minimum valid value of `(period)` is 256 µs.

In this section, we have gone through the analog pins available with a Micro:bit board. We also looked into the key functions that are used to operate various analog and digital devices. In the next section, we will explore the functionality of the potentiometer.

Potentiometers

The **potentiometer** (also known as **pot**) is a commonly used analog device. It is a three-pin device that offers variable resistance with the help of the middle pin. To connect it with Micro:bit, as depicted in *Figure 9.2*, we need to join the middle pin of a potentiometer to the analog pin and side pins to the power and ground:

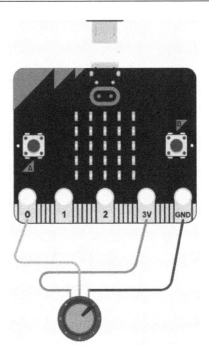

Figure 9.2 – Connecting a potentiometer to Micro:bit using P0

Once the connections are made, as shown in *Figure 9.2*, then the Micro:bit needs to be programmed, and the values from the potentiometer can be read for further use. As depicted in *Figure 9.2*, the potentiometer has three pins: one pin is connected to the ground and another is associated with the supply voltage; in the case of Micro:bit, it supplies 3 V of power. The middle pin is the point of consideration in the circuit connection, and it could be connected to any analog pin, such as pins 0, 2, 3, 4, and 10. The changes will be made in the program based on the analog pin assigned. For pin 0 as an analog input, the code will be as follows:

```
from microbit import *
pot_value=pin0.read_analog() # assign variable to read the pin0
value
print(pot_value)#the the values from pot
```

In the preceding code, pot_value is the variable assigned to collect the reading from the potentiometer connected to pin 0. In this program, we are importing the library of Micro:bit so that we can use the inbuilt functions of MicroPython. In the second line of the program, the pot_value variable is assigned to read the middle pin value of the potentiometer, and in the third line of the program, the print operation is performed. The values of a potential meter reading will be seen as integer numbers, which can be further converted into voltage values. We can use these output values for controlling various input-output operations. In *Figure 9.3*, the readings are directly coming from the potentiometer through the continuous rotation of the knob; for every movement of the knob, we will get just one value. *Figure 9.3* represents values based on multiple movements done with the help of the potentiometer's knob rotation:

```
Shell
   108
   125
   124
   102
   56
   20
   13
   14
   29
   94
   126
   125
   111
   49
   43
   44
   69
   127
   138
   138
```

Figure 9.3 – Readings from the potentiometer in a shell window

We can use these readings for calculation purposes as we turn the potentiometer from zero or low value to a high value; we can see the readings moving from 0 to 1,023, as observed by rotating the potentiometer from minimum to maximum and vice versa. These values can be used further to calculate the voltage across the potentiometer at any given position. The formula to calculate the value is as follows:

$$Voltage = Sensor\ reading\ \times \left(\frac{Supply\ Voltage}{1023}\right)$$

Here, the value is the integer we are getting while rotating the center tap of the potentiometer. For example, if we took the first reading from *Figure 9.3*, that is, 108, the voltage can be calculated as follows:

$$Voltage = 108 \times \left(\frac{3}{1023}\right) = 0.3V$$

If we take the highest value, which is 1,023, then the output voltage will be 3 V. In this way, we can use the potentiometer as a voltage divider. The voltage divider has many applications where voltage level adjustments are required.

The value of resistance can be calculated using the following:

$$resistance = Pot_{full\_resistance} \times \left(\frac{Value}{1023}\right)$$

For example, if the potentiometer value is 10,000, consider the first value from *Figure 9.3*, that is, 108, and put them in the preceding formula:

$$resistance = 10,000 \times \left(\frac{108}{1023}\right) = 1055.71\ Ohm$$

As shown in the code, the pot_value variable reads the integer values from the potentiometer:

```
from microbit import *
while True:
    pot_value=pin0.read_analog()
    print(pot_value)
    if pot_value <= 10:
        display.show(Image.SAD)
    else:
        display.show(Image.HAPPY)
    sleep(125)
```

Whereas the print function shows the integer value coming out of the pot based on the pot_value data, the Micro:bit will display happy and sad images.

In this section, we have gone through connecting a simple analog device, that is, the potentiometer, and operating it with the help of the program. We have also observed that the values fetched from the potentiometer can be printed on the shell window of the Thonny IDE and can also be converted into voltage and resistance values. In the next section, we will explore the application of photoresistors.

Photoresistors

A **photoresistor** is a device that is very sensitive to light or illuminance. Based on the presence of brightness in an ambient environment, the value of resistance can go up or down. This device helps control the circuits to switch on or off based on light sensitivity. For example, in the street light circuit, the lights should be turned on at night and turned off in the daytime. The working mechanism of a photoresistor is such that the resistance drops to a few Ohms when the illuminance is high and the resistance rises to higher values when the illuminance is low.

A photoresistor is also known as a **photocell**, **conductive photocell**, or **Light-Dependent Resistor** (**LDR**). In *Figure 9.4*, a symbolic representation of a photoresistor is shown:

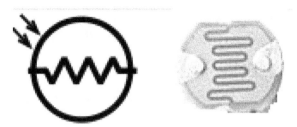

Figure 9.4 – A symbolic representation of a photoresistor with a light-sensitive surface (courtesy: https://en.wikipedia.org/wiki/Photoresistor)

On the left-hand side of the diagram, the two arrows indicate the illuminance or the light, whereas the resistor symbol in the circle indicates the variation of resistance due to light facing the photoresistor. The right-hand side of *Figure 9.4* represents an LDR with a sensitive surface through which light passes.

Figure 9.5 shows a connection of the photoresistor to the Micro:bit:

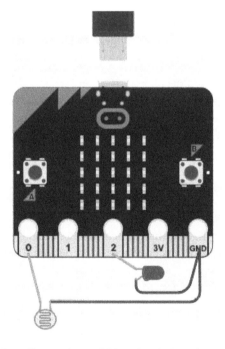

Figure 9.5 – Connecting an LDR and an LED to the micro: bit

A photoresistor is a two-terminal device – one of the terminals is connected to an analog pin, that is, pin 0, and another terminal is connected to the ground. Based on the photoresistor's value, the **Light-Emitting Diode** (LED) will change its state. After doing hardware configuration, we need to write a program so the value of a potentiometer can be reflected on a switching operation:

```
from microbit import *
x = pin0.read_analog()
print(x)
sleep(1000)
while True:
    y = pin0.read_analog()
    if y < x-50:# here 50 indicates sensitivity
        pin2.write_digital(1)
    else:
        pin2.write_digital(0)
        print(y)
```

In the preceding program, two variables are defined, that is, x and y. The x variable is for collecting the integer values coming from the photoresistor. y is implemented as a conditional statement to compare with the present value of x, which reflects the change in brightness. In the case that the value of the y variable is less than x, the LED will be turned on – otherwise, it will remain off.

In this section, we have gone through the photoresistor, and we have designed a circuit using a LED and photoresistor. With the help of the program, we also observed the on- and off-states of the LED. In the next section, we will cover PWM signals and their applications.

PWM signals

PWM is a way of handling analog write operations. It is a set of repetitive signal pulses, where the time of the pulse is in milliseconds, and the width of a pulse can be controlled using the `write_analog()` operation. When an analog device is connected to the Micro:bit, the read operation can be performed using `read_analog()`, and at the same time, when an output analog signal needs to be generated, it can be done using PWM. To do so, a duty cycle needs to be defined. The **duty cycle** can be explained as a measure of time in which a signal is active or a **General Purpose Input/Output** (GPIO) device receives an output. It can be easily understood with the help of this figure:

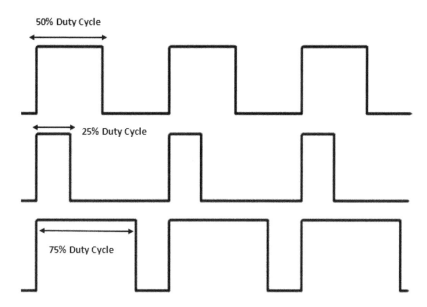

Figure 9.6 – PWM with the duty cycle (courtesy: https://microbit-micropython.readthedocs.io/_/downloads/hu/latest/pdf/)

In *Figure 9.6*, the first pulse indicates an equal distribution of high and low values; it implies that the output device connected to a particular pulse can stay on and off for the same amount of time. Therefore, the on-time and off-time are equally distributed, which is why the first pulse's duty cycle is 50%. Meanwhile, in signal two, the pulse remains on for less time and off for more time. In general, 25% of the time, it is on, and 75% of the time, it will remain off. As per the definition of the duty cycle, only the time when the signal is active is considered. In the last pulse, the device is on for most of the time and remains off for a concise amount of time. It can be said that for 75% of the time, the pulse is high, and for the remaining 25% of the time, the pulse is low; that is why the duty cycle, in this case, is 75%. Some devices require a certain level of continuous output values to operate; in these cases, we can use PWM signals. One example could be a servo motor.

Servo motors using PWM

A **servo motor** is a rotatory device with torque and velocity controlled through voltage and current. The main components in a servo motor are a driver circuit, which contains the power and the feedback loop, a potentiometer that regulates the voltage value, a **direct current** (**DC**) motor that performs a rotatory operation, and the gears to reduce the **revolutions per minute** (**RPM**) value so that the torque of a motor can be increased. *Figure 9.7* shows the connection of a servo motor with the Micro:bit:

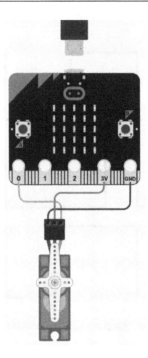

Figure 9.7 – Connecting a servo motor to the Micro:bit

As shown, the servo motor has three pins; two pins connect the ground and supply voltage to the Micro:bit, and the signal pin is attached to P0. Now, to observe the rotatory motions, we need to program the Micro:bit board accordingly:

```
from microbit import *

while True:
    pin0.write_analog(60)
    sleep(500)
    pin0.write_analog(80)
    sleep(500)
    pin0.write_analog(130)
    sleep(500)
```

In the preceding program, `pin0` is used as an analog output pin, passing an output pulse using a PWM signal through `pin0.write_analog(60)`, and the `sleep()` function is implemented to generate the delay. The program passes three different values using `write_analog()` to rotate the servo. In this way, a servo motor's precise motion is performed in three different steps. The values assigned to `write_analog()` can range from 0 to 1,023. The zero represents a 0% duty cycle and 1,023 represents a 100% duty cycle.

In this section, we have explored PWM signals with the help of working principles. We have also gone through key terms such as pulse and duty cycle. We have integrated the servo motor and have also gone through a simple program to run the motor. In the next section, we will look into connecting multiple analog devices to the Micro:bit board.

Handling multiple analog devices

After going through the *PWM signals* section, it is now time to connect more than one analog device to the Micro:bit. In this section, we will connect two analog devices, one as an input with `read_analog()` and another as an output device controlled with `write_analog()`. In *Figure 9.8*, at pin 0, a potentiometer is connected, and at pin 1, a servo motor is connected. Both devices are provided with a 3 V supply to complete the electrical circuit. To control the servo motor, a potentiometer passes the voltage levels acting as input values to the servo motor. The servo motor will rotate according to the input from the potentiometer:

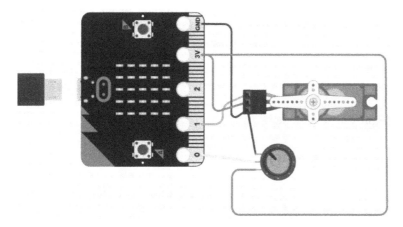

Figure 9.8 – Connecting the potentiometer as input device and the
servo motor as an output device to the Micro:bit

Figure 9.9 shows both a program and its output in the shell window:

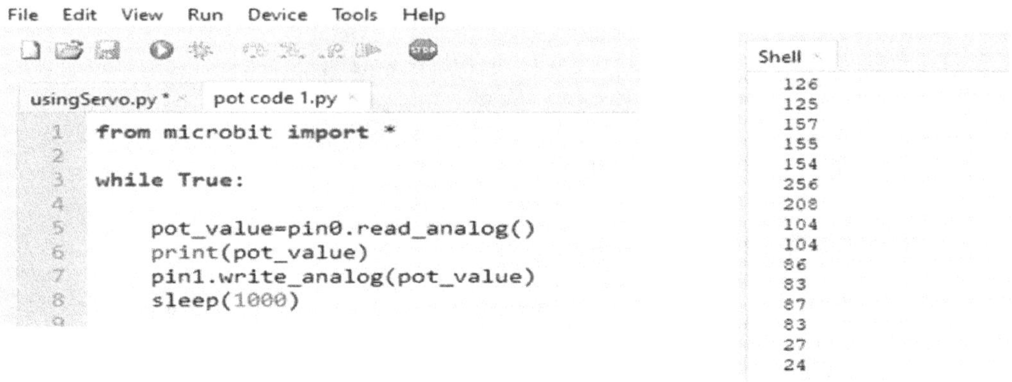

File   Edit   View   Run   Device   Tools   Help

usingServo.py *   pot code 1.py

```
1    from microbit import *
2
3    while True:
4
5        pot_value=pin0.read_analog()
6        print(pot_value)
7        pin1.write_analog(pot_value)
8        sleep(1000)
```

Shell

```
126
125
157
155
154
256
208
104
104
86
83
87
83
27
24
```

Figure 9.9 – Operating the servo motor through a potentiometer using the Micro:bit

Observing the program, the first line is for importing the library. A `while` loop is executed, reading the integer values at pin 0 coming from the potentiometer, which are then assigned to pin 1. Using the `write_analog()` operation, those values are given to the servo motor. Based on the integer value, different voltage levels are generated, which are provided as input to the servo motor. It generates a duty cycle for the servo motor, getting rotatory motions.

In this section, we have connected the potentiometer and the servo motor, and based on the value of the potentiometer, the servo motor moves. We have done this using the `read_analog` and `write_analog` functions. In the next section, we will explore the PWM application for a **red, green, and blue light-emitting diode (RGB LED)**.

PWM using an RGB LED

In the previous section, we observed the use of `write_analog()` for PWM signal generation. The same principle can be employed to achieve different color intensities through an RGB LED. An RGB LED differs from the usual LED in multiple ways; the first difference is the number of pins; in a standard LED, two pins are provided, whereas an RGB LED has four pins. In *Figure 9.10*, an RGB LED is depicted. It is visible that an RGB LED has red, green, and blue pins, which indicate the colors, and a positive pin connects to the ground:

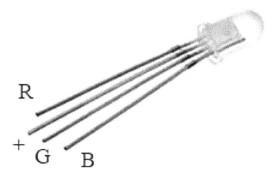

Figure 9.10 – An RGB LED (courtesy: https://en.wikipedia.org/wiki/Light-emitting_diode)

The most extended pin, as shown in *Figure 9.10*, indicates an anode. By providing the PWM signal, the LED will glow in various colors. Hence, an RGB LED is used in applications where several light hues are required from a single LED. As we have done in the case of handling the potentiometer and servo motor, we are now going to connect a potentiometer and an RGB LED together. In *Figure 9.11*, the hardware connection is demonstrated:

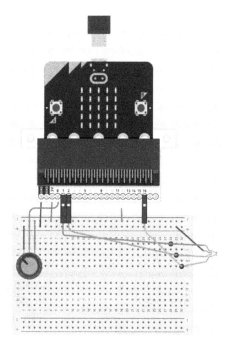

Figure 9.11 – Connecting an RGB LED and a potentiometer to the Micro:bit

A potentiometer and an RGB LED are connected to the Micro:bit. To connect the RGB LED, we need four pins. We can observe that pin numbers 1, 2, and 16 are connected to three terminals of the RGB LED, whereas, for the potentiometer, pin 0 is used. After making this combination, it is expected that as the potentiometer rotates, the LED will glow in different colors. With the help of the following program, we can control the LED glow. This is done by using potentiometer values and feeding them to the RGB LED:

```
from microbit import *
pot_val=pin0.read_analog()
print(pot_val)
while True:
    if pot_val <= 0:
        pin16.write_analog(pot_val)
    elif pot_val <= 400:
        pin1.write_analog(pot_val)
    elif pot_val <= 1000:
        pin2.write_analog(pot_val)
```

In the preceding program, the Micro:bit libraries are imported, and then a `pot_val` variable is assigned, which reads the value from the potentiometer. The exact value as printed using `print(pot_val)`. A few `if` and `elif` conditions are also provided, enabling the RGB pins connected in the circuit (depicted in *Figure 9.11*). Based on `pot_val`, pin numbers 16, 1, and 0 will be activated.

In this section, we have gone through the basics of an RGB LED and its pin diagram and have connected it to the Micro:bit board. The program helps control the LED glow with the range of potentiometer values fed to different pins of an RGB LED. In the next section, we will explore the joystick and its application.

Joysticks

A **joystick** is a very commonly known device in the gaming world. The working principle of a joystick is to monitor the variance on the *x*-axis and *y*-axis. In a conventional joystick, two potentiometers are connected on both the *x* and *y*-axes in 2D. The variation in the potential is observed with the help of a driver circuit. The initial position of a joystick is in the center. Based on the position of the joystick, the variance is calculated using the driver circuit. The driver circuit will calculate the value to locate its actual position – that is, up, down, right, and left. It also has a push button on the center position, which can be programmed. The joystick has five pins:

- **Ground (GND) and supply (VCC)**: Connected to the ground and the supply

- **VRx or Hor**: Reads the values from the *x* axis

- **VRy or Ver**: Reads the values from the *y* axis

- **SW**: The switch or push button

In *Figure 9.12*, the connection of a joystick can be observed:

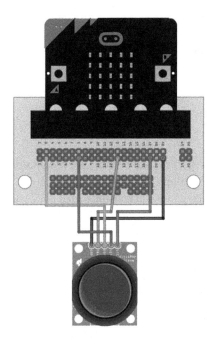

Figure 9.12 – Connecting a joystick to the Micro:bit board

The GND and VCC are connected to the respective pins of the Micro:bit. The VRy pin is connected to pin 1, whereas the VRx pin is connected to pin 0, and the switch pin is connected to pin 2 of the Micro:bit board.

In the following code, the Micro:bit pins are assigned to the joystick, and based on the values received from the read_analog functions, we display the position of the joystick using the LED matrix:

```
from microbit import *
pin0.read_analog()
pin1.read_analog()
pin2.read_digital()
pin2.set_pull(pin2.PULL_UP)
while True:
        value=pin0.read_analog()
        if value<300:
```

```
            display.show(Image.ARROW_W)
    elif value>600:
            display.show(Image.ARROW_E)
    value=pin1.read_analog()
    if value<300:
            display.show(Image.ARROW_N)
    elif value>600:
            display.show(Image.ARROW_S)
    value=pin2.read_digital()
    if value ==0:
            display.scroll("button")
```

In the preceding program, the *x* axis, *y* axis, and a button are programmed. Pin 2 is read_digital due to its usability for the switch functions. In pin2.set_pull(pin2.PULL_UP), set_pull enables a pull of the resistor connected to pin 2. It signifies that pin 2 will be enabled when the resistor value goes up. With analog pins, it is common to see the implementation of pull-up and pull-down resistors. In the while section of the code, we can observe that a value variable is assigned to check the analog values coming from pin 0, pin 1, and pin 2.

Figure 9.13 indicates the outcome of the operations performed using the joystick:

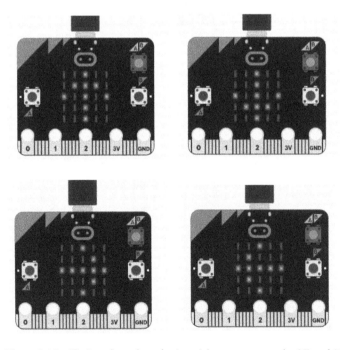

Figure 9.13 – Motions based on the joystick moments on the Micro:bit

Based on the movement of the joystick, we can see arrow signs on the Micro:bit.

In this section, we have gone through the working principle of a joystick. We have also connected it to the Micro:bit board and programmed it. We have used the LED display matrix to visualize the movement of the joystick.

Summary

In this chapter, we have explored the functioning of analog input-output devices with the help of the Micro:bit. We have demonstrated that using the built-in methods, `read_analog()` and `write_analog()`, we can print the input values and pass them to an analog device connected to the micro: bit. The functionality of the potentiometer as a variable resistor and voltage divider has been observed. We have completed the calculations required to convert the integer values from the potentiometer into the resistance and voltage levels. We have gone through the working principles and programming of the RGB LED, servomotor, photoresistor, and joystick.

In the next chapter, we will explore using more complex input-output devices such as accelerometers and compasses with the Micro:bit. We will design the program to explore the capabilities of these devices in real-life scenarios.

Further reading

You can find more information about the analog operations of the Micro:bit implementation of MicroPython at `https://microbit-micropython.readthedocs.io/en/v2-docs/pin.html` and `https://microbit-micropython.readthedocs.io/en/v2-docs/pin.html#pin-functions`.

Part 4:
Advanced Hardware Interfacing and Applications

You will explore how to interface and write programs for advanced hardware in this section. You will learn about the advanced built-in hardware features such as internal sensors and radio communication. You will also learn how to interface with external hardware displays and wearable electronics. There are many mini-projects for you to implement using these hardware components.

This section has the following chapters:

- *Chapter 10, Working with Acceleration and Direction*

- *Chapter 11, Working with NeoPixels and a MAX7219 Display*

- *Chapter 12, Producing Music and Speech*

- *Chapter 13, Networking and Radio*

- *Chapter 14, Advanced Features of the Micro:bit*

- *Chapter 15, Wearable Computing and More Programming Environments*

Working with Acceleration and Direction

In the previous chapter, we learned about the concepts of analog input and pulse-width modulation. We worked with analog devices such as potentiometers, photoresistors, and servo motors. We demonstrated their workings with the help of MicroPython examples in the previous chapter.

This chapter will explore the Micro:bit's onboard components, such as an accelerometer and compass. We will design examples to understand the functioning of these onboard components. We will investigate the method of data storage in a local drive. We will also explore how to plot graphs for the data coming through an accelerometer. The methods involved in this chapter will be essential to handling complex hardware and data patterns. The following is the list of topics we will investigate and demonstrate in this chapter:

- Accelerometers
- Data logging and plotting
- Compasses
- Combining audio with a compass

Let's explore them together.

Technical requirements

For this chapter, we will need the usual setup with a Micro:bit v2.

Accelerometer

An **accelerometer** is a device that measures the rate of change concerning the direction of force applied to it. It is a device that helps us to measure the vibrations, shock, tilt, and orientation in three dimensions. Now, the question may arise of how an accelerometer can carry out such measurements.

Let us consider an example of palm-top devices. Based on the position of the user, the orientation of devices changes. If a user has a tablet PC and uses it horizontally to have a more comprehensive view, the auto-rotate option helps change the view to horizontal. Users don't even need to make the change manually. This is because an accelerometer is working in real time and the change in orientation will be detected based on how the device is positioned. These features are also available in smartwatches (when you move your hand, the display will turn on), earpads (take them away from the ears and they automatically stop), and so on. Hence, accelerometers are now essential components of consumer electronics.

Let's look at the position of the accelerometer on the Micro:bit:

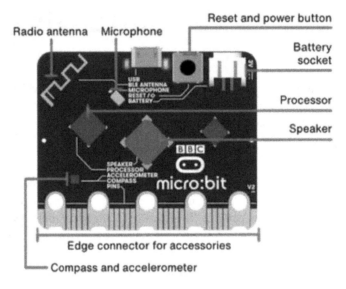

Figure 10.1 – Position of the accelerometer on the Micro:bit board (courtesy: https://microbit.org/get-started/user-guide/overview/)

The accelerometer is attached to the Micro:bit board as an onboard component with a compass. Although accelerometers and other similar devices can be attached to hardware using the GPIO pins available, in the case of the Micro:bit, the availability of the onboard devices makes it easy to use and less complex for developers:

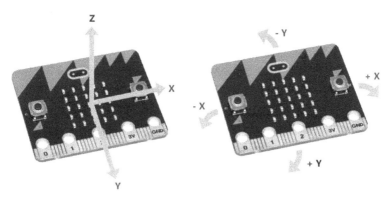

Figure 10.2 – Illustrating the accelerometer on a Micro:bit (courtesy: `https://microbit-challenges.readthedocs.io/en/latest/tutorials/accelerometer.html`)

As shown in *Figure 10.2*, on the tilt action in the **X**, **Y**, and **Z** directions, the displacement is calculated. The same can be viewed with the simple program demonstrated as follows: through the use of a built-in `accelerometer.get_x()` method, we get the acceleration value of the *x* axis, and the same can be done for the *y* and *z* axes too. The `print()` statement shows them in the shell window:

```
from microbit import *
while True:
    Tilt_x= accelerometer.get_x()
    print("Tilt_x ", Tilt_x)
    sleep(1000)
```

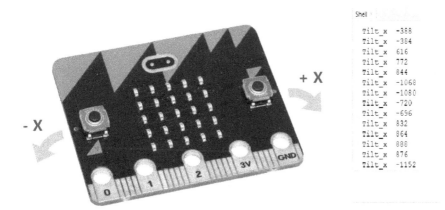

Figure 10.3 – Output of the program based on the tilt on the x axis

Figure 10.3 shows the output of the preceding program. As observed in the shell window, both positive and negative values are shown; the reason is also demonstrated in the Micro:bit diagram. Due to left and right movement on the *x* axis, both values will be visualized. If the accelerometer is tilted to the right, we will receive positive values, and negative values will be received when it is tilted to the left. The code, as mentioned earlier, can also be extended to obtain the value of the *y* and *z* axes.

In this section, we have explored the functionality of an accelerometer. We have also observed how to fetch the data from the accelerometer using the program. In the next section, we will see the data logging features and graph generation using an accelerometer.

Data logging

Figure 10.3 only shows the changes in the *x* axis data. Just one axis is insufficient for making effective decisions, such as fall detection, pitches, yaws, and rolls. We need to collect data from the *x*, *y*, and *z* axes. This data will make interpretation more accurate and reliable, as fall patterns can't be analyzed just from one axis. Hence, it is essential to have data stored in a file so it can be used for analysis. Data logging is a method of storing the data coming from sensors, actuators, or any subsystems to keep a record, which will further help decision-making and monitoring. Data visualization is equally essential; it is easier to observe the significant changes via graphs and figures.

In the following program, the movement of the Micro:bit board is recorded with the help of an accelerometer. Based on the tilt, we are displaying the image on an LED array display:

```
from microbit import *
while True:
    sleep(120)
    print(accelerometer.get_values())
    movement = accelerometer.current_gesture()
    if movement == "shake":
        display.show(Image.ANGRY)
    else:
        display.show(Image.HAPPY)
```

In the preceding program, `accelerometer.get_values()` fetches the values for the *x*, *y*, and *z* axes. Based on the input values, the **shake** motion is being executed; this means that when the user shakes the Micro:bit board, it will display an angry gesture, and the happy gesture will be displayed in a normal state.

For this particular program, we are using the Mu editor. The advantage of the Mu editor is that it can show the plot graph and the accelerometer's real-time values:

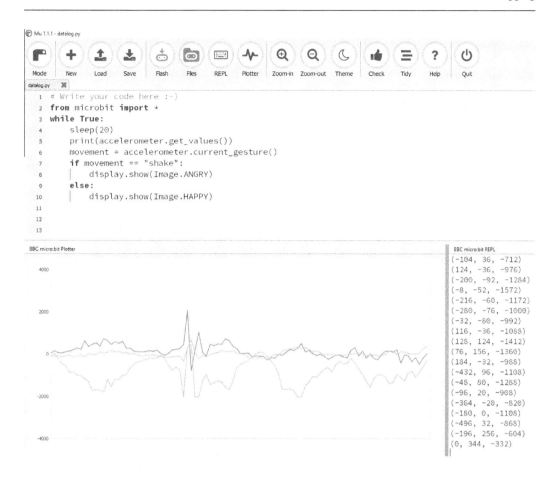

Figure 10.4 – Illustrating the functionality of the Mu editor

As shown in *Figure 10.4*, using the Mu editor, we can check the code functionality by clicking on the **Check** icon. Apart from that, the code can directly be flashed to the Micro:bit using the **Flash** icon. By clicking on **REPL**, it will show the real-time values of the accelerometer. The graph can be plotted using the **Plotter** icon shown on the toolbar. The three signals in the **Plotter** window represent the x, y, and z values from the accelerometer. The signals in the graph represent x, y, and z coordinates, ranging from positive and negative values based on the motion detected in real time. The data log is also available in the local repository.

As shown in *Figure 10.5*, to track the data location, we need to go to the Mu editor's local repository, that is, **mu_code**:

> MPL > mu_code ◀━━━━━━ 1

Name	Date modified	Type
data_capture ◀━ 2	8/10/2022 11:06 AM	File folder
fonts	7/19/2022 5:43 PM	File folder
images	7/19/2022 5:43 PM	File folder
music	7/19/2022 5:43 PM	File folder
sounds	7/19/2022 5:43 PM	File folder
static	7/19/2022 5:40 PM	File folder
templates	7/19/2022 5:39 PM	File folder

Go to mu_code folder and look for data capture folder

Figure 10.5 – Data log saved in the data_capture folder of mu_code

Inside the folder, we need to find the **data_capture** folder, as shown in *step 2* of *Figure 10.5*. Once we get inside the folder, the Excel file can be seen, as in *Figure 10.6*:

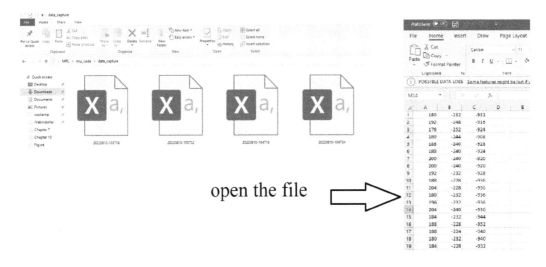

Figure 10.6 – Data saved as an Excel file and the values coming from the accelerometer

The Excel files are stored with a date and time stamp, so it would be easy for us to trace the data logs we are looking for. For example, in the **20220810-110618** filename, **20220810** represents the date in the YYYY-MM-DD format, and **110618** represents the time in the HH-MM-SS format.

In the preceding program, we have seen the usefulness of the Mu editor and have also gone through the data folders and files. In the next section, we will explore the compass and its application.

Compasses

Also known as a magnetometer, a **compass** measures the magnetic fields and is also an essential instrument for drones and aircraft to check the direction for navigation-related applications. In Micro:bit, as shown in *Figure 10.7*, the compass is mounted on the board. The functionality of the compass is such that it finds North at 0 degrees and 315 degrees as North West:

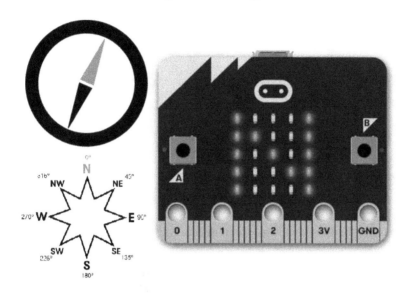

Figure 10.7 – Micro:bit with a compass, illustrating directions with degrees
(courtesy: https://microbit.org/projects/make-it-code-it/compass-north/)

From the figure, we can observe that South is at 180 degrees, East is at 90 degrees, and West is at 270 degrees. Hence, now, it is easier to track the direction using the compass, but first, we need to calibrate the compass. The compass.calibrate() function is used to perform the calibration:

```
from microbit import *
compass.calibrate()
```

In this process, the Micro:bit display panel will display the **Tilt to fill screen** message, scrolled onscreen, to rotate the board. Once it stops, for calibration, we need to tilt the Micro:bit board in all directions so that the LEDs on the Micro:bit board will be turned on. Please make sure that the program runs till all the LEDs are turned on, as shown in *Figure 10.8*:

Figure 10.8 – Display when the calibration LEDs are turned on

Once the calibration process is complete, a smiling face will appear for a short amount of time as shown in *Figure 10.9*:

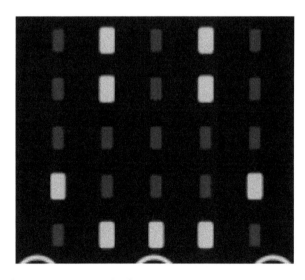

Figure 10.9 – After calibration, a smiley face appears, meaning the calibration is successful

After this, the screen will go blank, which signifies the process is finished, and the board is ready to track the direction in degrees through the heading:

```
from microbit import *
while True:
    print(compass.heading())
    sleep(300)
```

In the preceding program, the heading rotation will be displayed. The values will be given as 0 to 360 angles in degrees using compass.heading():

```
Shell
>>> %Run  -c  $EC
  359
  0
  359
  358
  351
  198
  281
  268
  291
  343
  114
  146
  127
```

Figure 10.10 – Display of the heading values from 0 to 359

Figure 10.10 shows the outcomes of the preceding program. 359 and 0 indicate North, 198 is close to South, and so on.

Similar to calibration and heading, we can also calculate the magnitude of the magnetic field around the Micro:bit:

```
from microbit import *
while True:
    print(compass.get_field_strength())
    sleep(300)
```

In the preceding code, `compass.get_field_strength()` indicates the values in **nanotesla (nT)**:

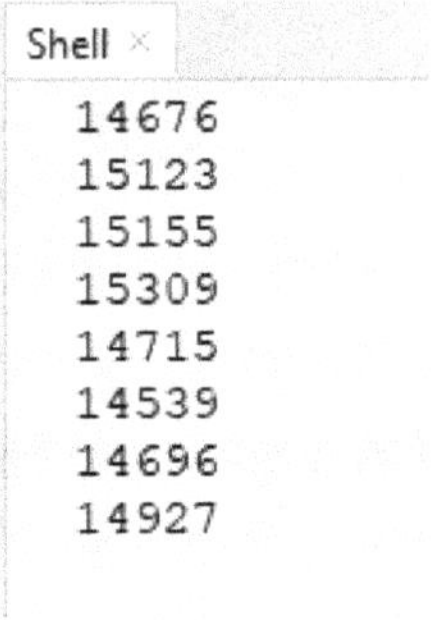

```
Shell ×
14676
15123
15155
15309
14715
14539
14696
14927
```

Figure 10.11 – Display of the output values of the magnitude of the magnetic field in nT

As indicated in *Figure 10.11*, the outputs can vary by variation in magnetic field strength.

Similar to observing magnetic field strength, we can also visualize the x, y, and z coordinate values of the compass. It can be done using the inbuilt `compass.get_` functions as seen in the following program:

```
from microbit import *
while True:
    degrees=compass.heading()
    magnitude= compass.get_field_strength()
    x_values = compass.get_x()
    y_values = compass.get_y()
    z_values = compass.get_z()
    print('degrees:{} magnitude:{} x_values:{} y_values:{}
z_values{}' .format(degrees, magnitude, x_values, y_values, z_
values))
    sleep(300)
```

In the program, we have assigned five variables as `degrees` for the heading, `magnitude` for the magnetic field strength, and `x_values`, `y_values`, and `z_values` for retrieving the values of the x, y, and z coordinates. `compass.get_y()` will provide the data for the y coordinates of the compass and the values for the x and z coordinates will also be retrieved using `compass.get_x()` and `compass.get_z()`. At the end of the code, the print will show the outcomes on the shell window, as shown in *Figure 10.11*:

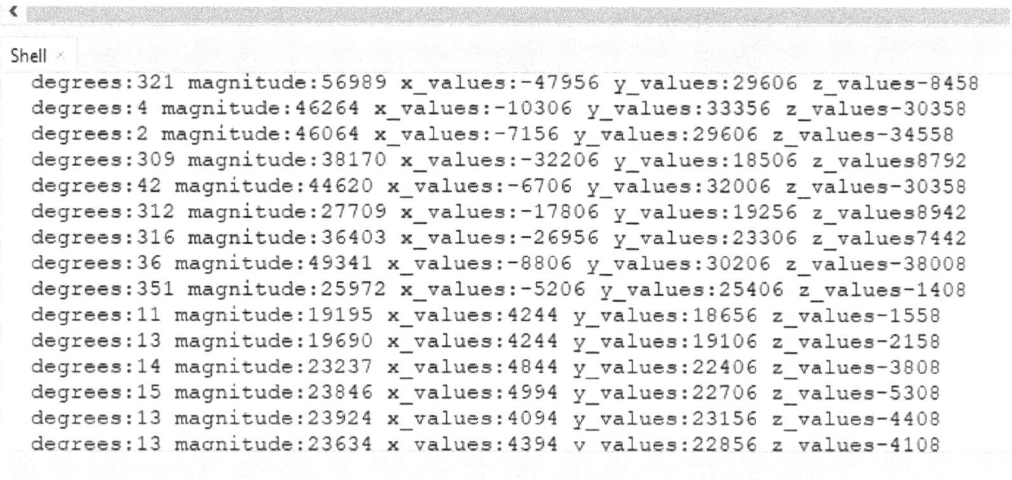

```
Shell
   degrees:321 magnitude:56989 x_values:-47956 y_values:29606 z_values-8458
   degrees:4 magnitude:46264 x_values:-10306 y_values:33356 z_values-30358
   degrees:2 magnitude:46064 x_values:-7156 y_values:29606 z_values-34558
   degrees:309 magnitude:38170 x_values:-32206 y_values:18506 z_values8792
   degrees:42 magnitude:44620 x_values:-6706 y_values:32006 z_values-30358
   degrees:312 magnitude:27709 x_values:-17806 y_values:19256 z_values8942
   degrees:316 magnitude:36403 x_values:-26956 y_values:23306 z_values7442
   degrees:36 magnitude:49341 x_values:-8806 y_values:30206 z_values-38008
   degrees:351 magnitude:25972 x_values:-5206 y_values:25406 z_values-1408
   degrees:11 magnitude:19195 x_values:4244 y_values:18656 z_values-1558
   degrees:13 magnitude:19690 x_values:4244 y_values:19106 z_values-2158
   degrees:14 magnitude:23237 x_values:4844 y_values:22406 z_values-3808
   degrees:15 magnitude:23846 x_values:4994 y_values:22706 z_values-5308
   degrees:13 magnitude:23924 x_values:4094 y_values:23156 z_values-4408
   degrees:13 magnitude:23634 x_values:4394 y_values:22856 z_values-4108
```

Figure 10.12 – Illustration of headings in degrees, the magnitude, and
the x, y, and z coordinate values in the shell window

In this section, we have gone through the basic features of a compass and its application. We have also observed with the help of a program that it can display headings and field strength for magnetic fields nearby. We have also been able to visualize the values from the compass, such as degrees, magnitudes, and *x*, *y*, and *z* axes values.

Audio and compass

This section will try to integrate the speech library and compass to assist with the direction information. The speech library will convert the heading into audio, and we have already observed the working of the heading function in the previous example. The code will be as follows:

```
from microbit import *
import speech
import music
while True:
    direction = compass.heading()
    print(direction)
    if direction < 35 :
        speech.say("North")
        display.scroll("N")
        music.play("E8:4")
    elif direction == 90:
```

```
        speech.say("East")
        display.scroll("E")
        music.play("D8:4")
    elif direction == 180:
        speech.say("South")
        display.scroll("S")
        music.play("C8:2")
    elif direction == 270:
        speech.say("West")
        display.scroll("W")
        music.play("B8:2")
```

Two libraries, music and speech, are imported into the preceding program. The music library will be used for playing the musical notes, whereas the sound library will play the speech. In the program, a variable name is initialized as a direction, which will collect the values for the heading information based on the heading data; North, South, East, and West directions will be recognized. After executing the program, when a user rotates the Micro:bit board, a specific tone and speech will be played once pointed in a specific direction.

Summary

In this chapter, we have explored the onboard components of the Micro:bit board, such as an accelerometer and a compass. Both are complex components with usage in real-life applications such as fall detection, shock analysis, and navigation. We have also explored data logging techniques using an accelerometer. We have also gone through compass programming using various methods. We also explored inbuilt functions and methods, such as accelerometer.get_values, compass. calibrate(), compass.heading(), and compass.get_field_strength().

In the next chapter, we will explore NeoPixel programming and interfacing with a MAX7219-based 8-digit Seven Segment display board.

Further reading

We can find more information about the accelerometer and compass in a Micro:bit implementation of MicroPython at https://microbit-micropython.readthedocs.io/en/v2-docs/ accelerometer.html and https://microbit-micropython.readthedocs.io/ en/v2-docs/compass.html.

11
Working with NeoPixels and a MAX7219 Display

In the previous chapter, we learned how to work with a built-in accelerometer and compass. We explored these onboard devices with MicroPython programming.

This chapter will explore how to interface NeoPixels and their compatible devices with the Micro:bit. We will also explore a MAX7219 driver-based display. We will learn how to program them and also create breathtaking projects with them. We will explore the following list of topics together:

- NeoPixel products
- The NeoPixel library
- Adding interactivity to the projects
- Interfacing a MAX7219/7221-based 7-segment 8-digit display

Let's get started with NeoPixel and MAX 7219 display programming.

Technical requirements

For this chapter, we will need any one of the following hardware products:

- NeoPixel or compatible strip or string
- NeoPixel or compatible ring
- NeoMatrix or compatible matrix
- MAX7219 7-segment 8-digit display

NeoPixel products

NeoPixels are a brand of a range of related products manufactured by Adafruit (`https://www.adafruit.com/category/168`). They are based on individual LEDs (WS2812, WS2811, and SK6812). These LEDs are **Red, Green, and Blue (RGB)** or **Red, Green, Blue, and White (RGBW)**-colored LEDs. They use the **Single Wire Protocol** for communication. Let's understand what it is. As you may recall, in *Chapter 6, Interfacing External LEDs*, we interfaced the Micro:bit board with RGB LEDs with four pins (one for each color and one for a common anode or common cathode). This circuit configuration can only support a limited number of RGB LEDs with the Micro:bit (or any board/device for that matter) as it has a limited number of GPIO pins that can be used for digital I/O. So, the drawbacks of this configuration are as follows:

- Utilization of too many pins

- Cannot support more RGB LEDs

- Must write separate code blocks to handle each LED

NeoPixel only uses one GPIO pin for many RGB LEDs. Yes! You read that correctly. It requires a single digital I/O pin of the board it is interfaced to. Depending on the configuration, a NeoPixel product can have any number of LEDs; we will have a look at several variations in this chapter, and learn how to build circuits and how to program them.

You can check online marketplaces or the product page of Adafruit (`https://www.adafruit.com/category/168`) to procure NeoPixel products. Alternatively, many manufacturers manufacture WS2812, WS2811, and SK6812-based products and sell them under their own brands. While purchasing, make sure that the product is **NeoPixel-compatible**.

The following figure shows a WS2812-based individual NeoPixel:

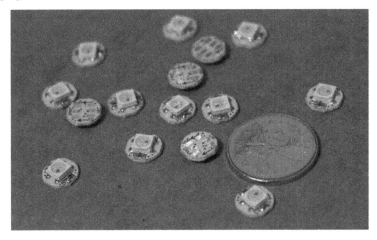

Figure 11.1 – Individual NeoPixel boards (courtesy: https://commons.
wikimedia.org/wiki/File:Mini_NeoPixel.jpg)

The following figure shows WS2812-based LEDs glowing in different colors. This should give you a fair idea of this hardware component's color spectrum and luminary capabilities:

Figure 11.2 – NeoPixel LEDs glowing in different colors (courtesy:
https://commons.wikimedia.org/wiki/File:WS2812Closeup2.jpg)

Let's discuss what pins the NeoPixel products typically have:

- **+ or DC pin**: This could be labeled as +5 DC, 4-7 DC, or V_{DD} on various products. We connect it to an external power supply of compatible output. The Micro:bit board may not be able to power the NeoPixel device if it has got too many LEDs. Some NeoPixel devices have more than one pin labeled +. One is for input voltage, while the other acts as power output if we plan to attach multiple NeoPixels in series. However, when using the series configuration, I prefer to power every NeoPixel product directly from the breadboard's power supply.

- **GND**: This one is to be connected to the common ground. If there is a pair of them in your NeoPixel product, the other GND pin should be connected to the GND pin of the Micro:bit. This provides the common ground for all the devices as all the GND pins of any NeoPixel or compatible product are connected internally.

- **D_{IN}**: This pin is the data input for the Single Wire Protocol. It is connected to one of the digital GPIO pins of the Micro:bit. Throughout this chapter, I will be using pin 0 for the demonstrations. However, you can use any pin.

- **D_{OUT}**: This is the data output pin, which is to be connected to the D_{IN} pin of the next NeoPixel product if we are connecting multiple NeoPixels in a series. Many NeoPixel-compatible products do not have this, and cannot be used for series configuration.

Now, let's have a look at the various NeoPixel form factor products one by one. I am using the Fritzing parts downloaded from the following URLs:

- `https://forum.fritzing.org/t/ws2812-rgb-led-strip-matrix/6339`
- `https://github.com/adafruit/Fritzing-Library`
- `https://github.com/vemeT5ak/fritzing-and-eagle/blob/master/SMD%20RGB%20LED%20(WS2812B).fzpz`

We have already seen figures of individual NeoPixel boards with a single LED. We can string them together in a serial chain, as follows:

Figure 11.3 – WS2821-based individual NeoPixel boards connected in a series

Connect all the V_{DD} pins to the external +5 V power supply and the VSS pins to the common ground. This will complete the circuit. So long as we provide enough external power, a reasonable number of NeoPixels can be attached without any issue.

We also have NeoPixels in the form of sticks, as follows:

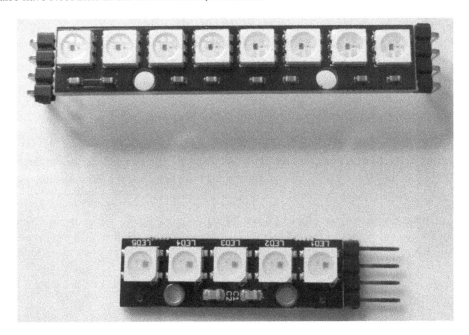

Figure 11.4 – NeoPixel sticks

We can see the pin labeling on the rear, as follows:

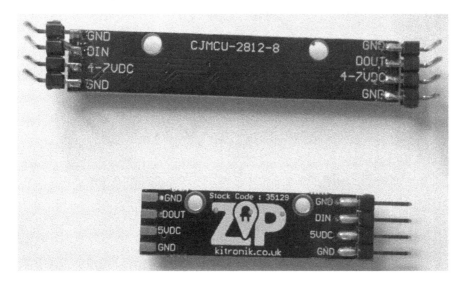

Figure 11.5 – NeoPixel sticks – pin description

Both these products have two sets of ground pins. Connect the +DC pin to the power supply for adequate output. One ground pin is to be connected to the Micro:bit's GND pin, while another ground pin is to be connected to the common ground. We can connect the D_{IN} pin to pin 0 of the Micro:bit as follows:

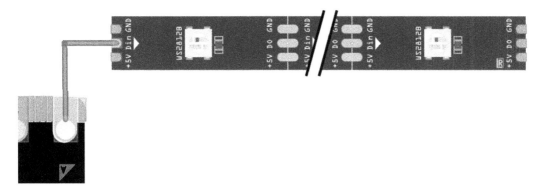

Figure 11.6 – NeoPixel stick/strip used in a circuit

Another popular product is the NeoPixel ring. The following figure shows the front of a ring with 16 LEDs:

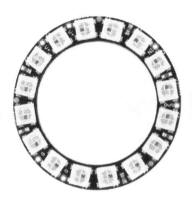

Figure 11.7 – The front of a 16-LED NeoPixel ring (courtesy: https://
commons.wikimedia.org/wiki/File:12664-02a.jpg)

The rear of this ring has pins, as shown here:

Figure 11.8 – The rear of a 16-LED NeoPixel ring (courtesy: https://
commons.wikimedia.org/wiki/File:12664-03a.jpg)

We can even have bigger rings, as shown here:

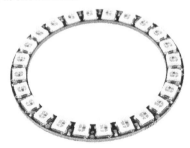

Figure 11.9 – A 24-LED ring (courtesy: https://commons.wikimedia.org/wiki/File:12664-03a.jpg)

We can create a circuit as follows:

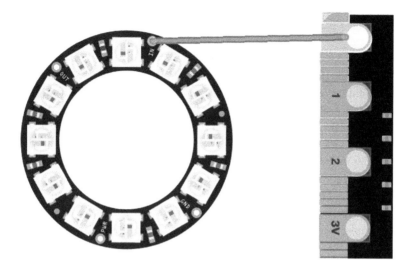

Figure 11.10 – A 12-LED ring connected to pin 0 of the Micro:bit

Do not forget to connect the PWR pin of the ring to +5V and the GND pin to a common ground. There are other form factors too. You can check them out at https://learn.adafruit.com/adafruit-neopixel-uberguide/form-factors.

We will be using this circuit for demonstration purposes; however, you can use any product.

In the next section, we will learn how to create interesting projects with the NeoPixel library.

The NeoPixel library

The NeoPixel library in the Micro:bit allows us to write programs for NeoPixel products and their compatible devices. You can find out more about the library at https://microbit-micropython.readthedocs.io/en/v1.0.1/neopixel.html. The library comes with MicroPython; we do not have to do anything extra to enable it. Let's start coding. I am going to connect the NeoPixel 12-LED ring to pin 0 for this chapter. The code examples have been written while considering this circuit. However, I will explain where to make changes if you have opted for a different configuration.

Check the following code example:

```
from microbit import *
from neopixel import NeoPixel
num_pixels = 12
ring = NeoPixel(pin0, num_pixels)
for i in range(0, 120, 10):
```

```
        print(i)
        ring[int(i/10)] = [0, 0, i]
    ring.show()
```

The first two lines are used to import the required libraries. Then, we assign the number of LEDs in the NeoPixel product to the num_pixels variable in the program. Then, we create a ring object for the NeoPixel product. We pass the pin number and the number of pixels as arguments to the constructor NeoPixel(). It is not mandatory to name the object ring in the program. It is just a variable name, and I usually name them to reflect real-life objects. If I were using a NeoPixel stick, then I would name it stick. Once the ring/stick object has been created, we can treat it as a list. If there are n pixels in the NeoPixel product, then the index of the list will start at 0 and end at n-1. We can assign a list representing the intensities of red, green, and blue to every member of the ring object. I have used a for loop for this. We can use various combinations of colors to achieve different effects. In the end, we can call the show() method to push all these values to the NeoPixel product. In this example, we will assign values ranging from 0 to 110 in increments of 10 to all 12 pixels (0, 10, 20… up to 110). This will create a static fading effect. We can modify the contents of the for loop as follows to create the blending effect:

```
for i in range(0, 120, 10):
    print(i)
    ring[int(i/10)] = [120-i, 0, i]
```

The print() call in the loops is only for debugging; you can comment it out if you wish.

We can create a rotating pattern with the following program:

```
from microbit import *
from neopixel import NeoPixel
num_pixels = 12
ring = NeoPixel(pin0, num_pixels)
shift = 0
values = [0, 10, 20, 30, 40, 50, 60, 70, 80, 90, 100, 110]
try:
    while True:
        for i in range(0, num_pixels, 1):
            curr_value = values[int((i + shift) % num_pixels)]
            ring[i] = [0, 110-curr_value, curr_value ]
        ring.show()
        sleep(1000)
        shift = shift + 1
```

```
except KeyboardInterrupt as e:
    print("Interrupted by the user...")
```

In the preceding program, we are defining a list known as values. We have written an infinite loop to assign the values of this list to the LEDs in the NeoPixel product in a rolling window fashion. In the infinite loop, we shift the window by one position in the list with the shift = shift + 1 statement. We can change the speed of the effect by modifying the argument that's passed to the sleep() call. We can see the printed values of the rolling window in the REPL shell.

By changing the values list, we can implement a lot of effects, as follows:

```
from microbit import *
from neopixel import NeoPixel
num_pixels = 12
ring = NeoPixel(pin0, num_pixels)
shift = 0
values = [[0, 0, 0], [0, 0, 120], [0, 120, 0], [120, 0, 0],
         [120, 120, 120], [120, 120, 0], [120, 0, 120], [0,
120, 120],
         [0, 0, 0], [0, 0, 40], [0, 40, 0], [40, 0, 0]]
try:
    while True:
        for i in range(0, num_pixels, 1):
            print(values[int((i + shift) % num_pixels)])
            ring[i] = values[int((i + shift) % num_pixels)]
        print('--')
        ring.show()
        sleep(1000)
        shift = shift + 1
except KeyboardInterrupt as e:
    print("Interrupted by the user...")
```

Run the preceding program to see the output. We will see another effect with the following list:

```
values = [[0, 0, 0], [0, 63, 0],
         [0, 0, 0], [63, 0, 0],
         [0, 0, 0], [0, 0, 63],
         [0, 0, 0], [0, 63, 0],
         [0, 0, 0], [63, 0, 0],
         [0, 0, 0], [0, 0, 63]]
```

We can create a chaser effect as follows:

```
from microbit import *
from neopixel import NeoPixel
num_pixels = 12
ring = NeoPixel(pin0, num_pixels)
try:
    while True:
        for i in range(0, num_pixels):
            ring[i] = [0, 0, 100]
            ring[i-1] = [0, 100, 0]
            ring[i-2] = [100, 0, 0]
            ring.show()
            sleep(50)
            ring.clear()
except KeyboardInterrupt as e:
    print("Interrupted by the user...")
```

The clear() method clears all the pixels when called. In this example, we have assigned different intensities of red, green, and blue to consecutive pixels to create a chaser effect. In each iteration, we change the index of those pixels. So, at any given moment, only three LEDs will be on and the rest of the LEDs in the ring will be off.

We can create a chaser effect with a single LED blinking with a random color by modifying the loop in the preceding code like so:

```
try:
    while True:
        for i in range(0, len(ring)):
            red = randint(0, 60)
            green = randint(0, 60)
            blue = randint(0, 60)
            ring[i-1] = [0, 0, 0]
            ring[i] = [red, green, blue]
            ring.show()

            sleep(100)
except KeyboardInterrupt as e:
    print("Interrupted by the user...")
```

In this example, we are randomizing the color of a single LED in every iteration of the loop.

We can also create beautiful LED patterns of red, green, and blue that glow in an infinite loop, as follows:

```
from microbit import *
from neopixel import NeoPixel
num_pixels = 12
ring = NeoPixel(pin0, num_pixels)
red = [31, 0, 0]
green = [0, 31, 0]
blue = [0, 0, 31]
delay = 500
def glowAll(col):
    for i in range(0, len(ring)):
        ring[i] = col
    ring.show()
    return 0
try:
    while True:
        glowAll(red)
        sleep(500)
        glowAll(green)
        sleep(500)
        glowAll(blue)
        sleep(500)
except KeyboardInterrupt as e:
    print("Interrupted by the user...")
```

The user-defined glowAll() function assigns the passed color to every LED in the NeoPixel product.

We can assign a random color to each pixel in each iteration, as follows:

```
from microbit import *
from random import randint
from neopixel import NeoPixel
num_pixels = 12
ring = NeoPixel(pin0, num_pixels)
delay = 500
def glowAll():
```

```
    for i in range(0, len(ring)):
        ring[i] = [randint(0, 255),
                    randint(0, 255),
                    randint(0, 255)]
    ring.show()
    return 0
try:
    while True:
        glowAll()
        sleep(500)
except KeyboardInterrupt as e:
    print("Interrupted by the user...")
```

To assign the same color to all the LEDs randomly, we have to modify the user-defined function, as follows:

```
def glowAll():
    red = randint(0, 255)
    green = randint(0, 255)
    blue = randint(0, 255)
    for i in range(0, len(ring)):
        ring[i] = [red, green, blue]
    ring.show()
    return 0
```

Try this modification and run the code to see the output. Do not forget to save the file with a different filename.

We can also create a beautiful fading effect with the following code:

```
from microbit import *
from neopixel import NeoPixel
num_pixels = 12
ring = NeoPixel(pin0, num_pixels)
delay = 1
step = 1
def glowAll(col):
    for i in range(0, len(ring)):
```

```
            ring[i] = col
        ring.show()
        return 0
try:
    while True:
        for i in range(0, 255, step):
            glowAll((i, 0, 0))
            sleep(delay)
        for i in range(255, 0, -step):
            glowAll((i, 0, 0))
            sleep(delay)
        for i in range(0, 255, step):
            glowAll((0, i, 0))
            sleep(delay)
        for i in range(255, 0, -step):
            glowAll((0, i, 0))
            sleep(delay)
        for i in range(0, 255, step):
            glowAll((0, 0, i))
            sleep(delay)
        for i in range(255, 0, -step):
            glowAll((0, 0, i))
            sleep(delay)
except KeyboardInterrupt as e:
    print("Interrupted by the user...")
```

We can adjust the delay and step variables to modify the fade's duration.

The MicroPython reference (https://docs.micropython.org/en/latest/esp8266/tutorial/neopixel.html) contains a code example for ESP8266 boards. It won't run on a Micro:bit device as is, so I modified it and made it run on the Micro:bit. The following code has been adapted for the Micro:bit. Let's import the libraries and initialize the variables:

```
from microbit import *
from neopixel import NeoPixel
num_pixels = 12
ring = NeoPixel(pin0, num_pixels)
```

Now, let's define the demo function:

```
def demo(ring, num_pixels):
    n = num_pixels
```

The following code block creates the cycle effect:

```
    # cycle
    for i in range(4 * n):
        for j in range(n):
            ring[j] = [0, 0, 0]
        ring[i % n] = [255, 255, 255]
        ring.show()
        sleep(25)
```

The following code creates the bounce effect:

```
    # bounce
    for i in range(4 * n):
        for j in range(n):
            ring[j] = [0, 0, 128]
        if (i // n) % 2 == 0:
            ring[i % n] = [0, 0, 0]
        else:
            ring[n - 1 - (i % n)] = [0, 0, 0]
        ring.show()
        sleep(60)
```

The following code creates the fade effect:

```
    # fade in/out
    for i in range(0, 4 * 256, 8):
        for j in range(n):
            if (i // 256) % 2 == 0:
                val = i & 0xff
            else:
                val = 255 - (i & 0xff)
            ring[j] = [val, 0, 0]
```

```
        ring.show()

    ring.clear()
```

Now, let's write the code that will call the function we just wrote:

```
try:
    while True:
        demo(ring, num_pixels)
except KeyboardInterrupt as e:
    print("Interrupted by the user...")
```

Since we are writing a separate function and passing it to the object of the NeoPixel product, extending this code for multiple NeoPixels connected to separate pins is easy. This program combines all the demonstrations we have seen earlier into a single program. You will find plenty of examples of NeoPixels for other boards. You can use the logic and the code provided here and then adapt it for your Micro:bit, as I did in this example, to hone your programming skills further.

We can also make a single LED glow in a random color in each iteration, as follows:

```
from microbit import *
from neopixel import NeoPixel
import random
num_pixels = 12
ring = NeoPixel(pin0, num_pixels)
try:
    while True:
        ring[(random.randint(0, num_pixels-1))] = \   [(random.
randint(0, 32)),
    (random. randint(0, 32)),
    (random.randint(0, 32))]
        ring.show()
        sleep(200)
        ring.clear()
except KeyboardInterrupt as e:
    print("Interrupted by the user...")
```

We can combine all the knowledge we have gained to create a rainbow effect, as follows:

```
from microbit import *
from neopixel import NeoPixel
num_pixels = 12
ring = NeoPixel(pin0, num_pixels)
def rainbow(np, num, offset, bright=1):
    rb = [(127, 0, 0), (127, 63, 0), (127, 127, 0), (0, 127,
0),
          (0, 127, 127), (0, 0, 127), (63, 0, 127), (127, 0,
0)]
    for i in range(num):
        t = 7*i/num
        t0 = int(t)
        r = round((rb[t0][0] + (t-t0)*(rb[t0+1][0]-rb[t0]
[0]))*bright) >> 2
        g = round((rb[t0][1] + (t-t0)*(rb[t0+1][1]-rb[t0]
[1]))*bright) >> 2
        b = round((rb[t0][2] + (t-t0)*(rb[t0+1][2]-rb[t0]
[2]))*bright) >> 2
        ring[(i+offset)%num] = [r, g, b]
n = 0
try:
    while True:
        rainbow(ring, num_pixels, offset = n)
        ring.show()
        n = n + 1
        sleep(200)
except KeyboardInterrupt as e:
    print("Interrupted by the user...")
```

In this example, we are defining a custom function called rainbow() that accepts the NeoPixel object, the number of pixels, the offset (we will explain this soon), and the brightness as arguments. Since this is a modular function, we can call it multiple times for different NeoPixels by changing the arguments. We have a predefined list of color values, rb, which is used to generate the rainbow colors. Based on the number of pixels passed to the function call, it computes the shades of the rainbow. We can control the brightness by passing the desired value as an argument. We are using the left-shift operator (>>) to shift the computed values of red, green, and blue toward the right by effectively reducing their brightness. The >>2 operation shifts a value to the right and reduces its magnitude. If you feel that the colors are too bright, you can change them to >>3 and >>4. The offset variable increments in every iteration and is passed to the function call. We use it to shift the value of the rainbow by one position every time. It creates the rotation effect. Run the code and see it in action. If you wish to see the changing values of color, you can print them with print().

This is how we can create great projects with NeoPixel components and Micro:bit. There is a lot of room to add your own creative ideas to these projects. As an exercise, create new lighting patterns with the same setup.

Adding interactivity to the projects

We can use the built-in push buttons **A** and **B** of the Micro:bit to control the speed of the rainbow. We have to use the if statement in the while loop. We need to have an extra variable for the speed as follows:

```
num_pixels = 12
ring = NeoPixel(pin0, num_pixels)
maxdelay = 400
mindelay = 10
delay = int(maxdelay+mindelay)/2
```

And we have to make the following modification to the loop:

```
try:
    while True:
        rainbow(ring, num_pixels, offset = n)
        ring.show()
        n = n + 1
        if (button_a.is_pressed() and (delay < maxdelay)):
            delay = delay + 10
        elif (button_b.is_pressed() and (delay > mindelay)):
            delay = delay - 10
        else:
            pass
        print(delay)
        sleep(delay)
except KeyboardInterrupt as e:
    print("Interrupted by the user...")
```

Now, run the program and control the speed of the rotation of the rainbow with the built-in buttons. You can also add additional logic in the `else` block if you wish.

Interfacing a MAX7219/7221-based 7-segment 8-digit display

MAX7219/7221 are 7-segment 8-digit driver ICs that use the serial interface. We can find their documentation at the following web pages:

- `https://www.maximintegrated.com/en/products/power/display-power-control/MAX7219.html`

- `https://datasheets.maximintegrated.com/en/ds/MAX7219-MAX7221.pdf`

Following is the photograph of a MAX7219-based 7-segment 8-digit LED display:

Figure 11.11 – 7-segment 8-digit display with pins

We can see all the pins clearly in *Figure 11.11*. Connect V_{CC} to the 3V pin of the Micro:bit and GND to the ground. Connect D_{IN} and CLK to pins 15 and 13 of the Micro:bit, respectively. Finally, connect CS to pin 0. We can use any digital I/O pin of the Micro:bit. I have written the demonstration programs assuming that pin 0 is connected to CS. If you decide to change the pin, please make the appropriate changes to the program.

I have referred to the library at `https://github.com/microbit-playground/matrix7seg` to interface the display with the Micro:bit. Download the file at `https://github.com/microbit-playground/matrix7seg/blob/master/matrix7seg.py` to your local computer. Now, open that file with the Thonny editor. To do so, in the main menu, go to **View** | **Files**:

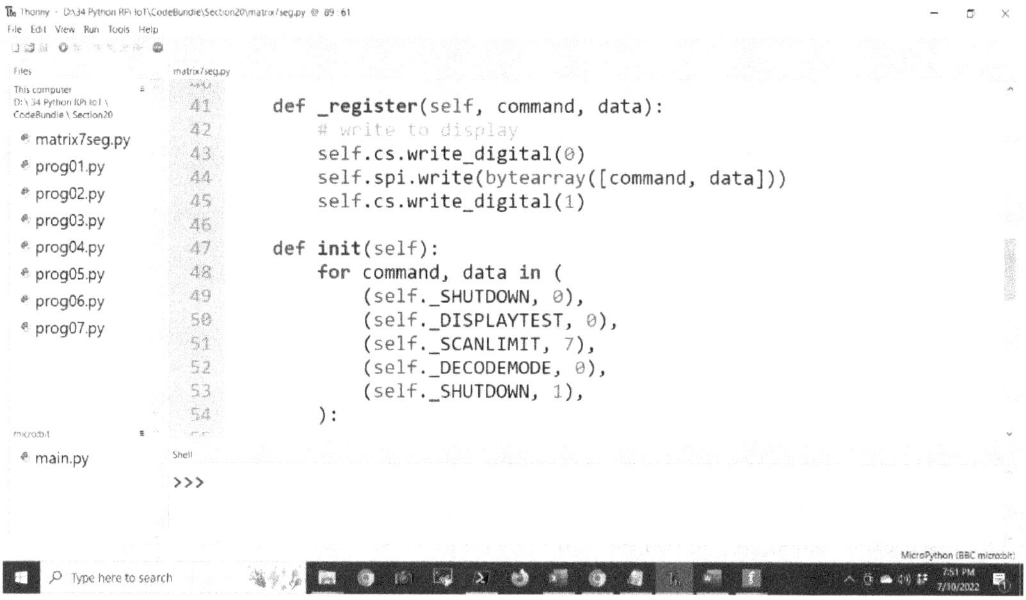

Figure 11.12 – Files view of Thonny

Make sure that you are not running any program in the REPL shell when you open this view; otherwise, the Micro:bit files will not be visible. In *Figure 11.12*, we can see that the Micro:bit has a main.py file. Now, save the current file (matrix7seg.py) to the Micro:bit. After we save the file to the Micro:bit, the Micro:bit file list will change, as follows:

micro:bit

Figure 11.13 – Files on the Micro:bit

This is how we can install a custom MicroPython library to any device running MicroPython. We can use the functionality from this library with the import statement. Save the following program to the Micro:bit as main.py:

```python
from matrix7seg import Matrix7seg
from microbit import spi, pin0, sleep
seg_display = Matrix7seg(spi, pin0)
seg_display.write_number(1234)
```

```
seg_display.show()
sleep(2000)
seg_display.write_number(1234, zeroPad=True)
seg_display.show()
sleep(2000)
seg_display.write_number(12345678)
seg_display.show()
sleep(2000)
seg_display.write_number(1234, leftJustify=True)
seg_display.show()
sleep(2000)
```

In the first two statements, we have imported all the needed modules. The `seg_display = Matrix7seg(spi, pin0)` statement creates an object for the display. The `write_number()` method is self-explanatory. We can pad the number with preceding zeros to fill the display and left-justify the text with the arguments. All the possible usages are shown in this example.

Let's create a simple counter, as follows:

```
from matrix7seg import Matrix7seg
from microbit import spi, pin0, sleep
seg_display = Matrix7seg(spi, pin0)
i = 0
try:
    while True:
        seg_display.write_number(i)
        seg_display.show()
        sleep(1000)
        i = i + 1
        if i > 9:
            i = 0
except exception:
    print("Interrupted by user...")
```

Here, we are displaying the value of an integer variable and incrementing it. When the value reaches 10, we reset it to 0. We can extend this program and add interactivity with the built-in push buttons. As an exercise, create a stopwatch using the display and the built-in push buttons. One push button should reset the stopwatch, while the other should stop it. You just need to modify the preceding example.

This is how we can use an external library in our code example for the Micro:bit.

Summary

This chapter explored the various NeoPixel products and the MAX7219-based 7-segment 8-digit display. We programmed them with the NeoPixel product and a custom library, respectively. As an exercise, try combining these LED-based displays with the built-in or external push buttons so that you have interactivity in your projects.

The code bundle for this chapter contains a few additional programs. As an exercise, go through those programs and understand their logic.

The next chapter will explore programming the Micro:bit to produce music and speech. We will learn how to connect an external speaker to the Micro:bit. We will also learn how to use a built-in microphone and speaker in the Micro:bit V2.

Further reading

The content available at the following URLs will enhance your understanding of NeoPixels further:

- `https://learn.adafruit.com/micro-bit-lesson-3-neopixels-with-micro-bit`

- `https://docs.micropython.org/en/latest/esp8266/tutorial/neopixel.html`

- `https://microbit-micropython.readthedocs.io/en/v1.0.1/neopixel.html`

- `https://www.adafruit.com/category/168`

If you plan to use any of the examples provided in the online tutorials, make sure that it is written for the Micro:bit, not any other board. If it is written for another board, then modify it appropriately.

12
Producing Music and Speech

In the previous chapter, we learned how to work with NeoPixel rings and strips. We also explored how to program the MAX7219-based 7-segment 8-digit display board. We worked with an external MicroPython library for the MAX7219 project and now we are comfortable working with external libraries. We created interesting projects with these cool external displays.

This chapter will explore a new dimension of the Micro:bit by discussing various methods of generating music and speech using Micro:bit. As discussed in the introductory chapters, the Micro:bit V2 has an inbuilt speaker. The programming module comes with inbuilt functions to generate tones and melodies. These can be easily heard using the inbuilt speakers. If we want to hear them louder, we can also connect to any external speaker. In this chapter, we will cover the step-by-step process of dealing with various methods of producing music and speech.

We will explore the following list of topics together:

- Connecting a speaker
- Melodies
- Custom melodies
- Tempo and pitch
- Working with a microphone
- Working with speech

Let us get started with music and speech with MicroPython and Micro:bit.

Technical requirements

Apart from the usual setup, the demonstrations in this chapter need the following components (which are not required for Micro:bit V2 users):

- Piezo buzzer
- Headphones/earphones with a 3.5 mm jack

Connecting a speaker

Speakers are electromagnetic components used for generating sounds. The internal circuit diagram of a speaker is shown in *Figure 12.1*. A speaker uses a thin magnetic membrane that can be excited by the electromagnetic field generated by the electric signals created by sound. As shown in the following figure, the North and South poles of the magnet will attract different parts of the membrane and generate different sounds. The faster or slower this membrane vibrates, the higher or lower the frequency of the output sound will be:

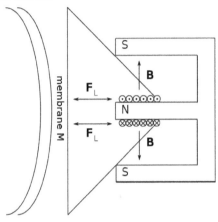

Figure 12.1 – Internal circuit diagram of a speaker (courtesy: https://upload.wikimedia.
org/wikipedia/commons/4/45/Simplified_loudspeaker_diagram_with_forces.svg)

The Micro:bit V2 has an inbuilt speaker, which is located on the backside of the board. *Figure 12.2* shows the location of the inbuilt speaker on the Micro:bit:

Figure 12.2 – Location of the in-built speaker on the Micro:bit V2 (courtesy: https://
microbit-micropython.readthedocs.io/en/v2-docs/_images/speaker.png)

In the Micro:bit V1, the inbuilt speaker is not available. Therefore, we can only rely on external speakers or headphones to hear sounds generated by the Micro:bit. We can also connect a speaker or headphones to the Micro:bit V2. The process of connecting the speaker to Micro:bit is the same, irrespective of the version. A speaker has two terminals – that is, *positive* and *negative*. The negative terminal should be connected to the ground pin of the Micro:bit, and the positive pin can be connected to either of the round pins (0, 1, or 2). A headphone with a 3.5mm jack also has the same two terminals on the jack. *Figure 12.3* shows the connection between a headphone jack and the Micro:bit:

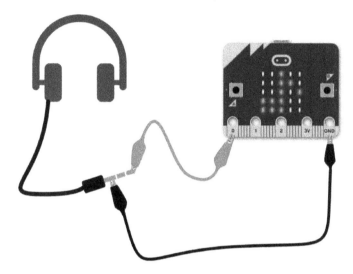

Figure 12.3 – Connecting a headphone/speaker to the Micro:bit

In this section, we discussed how to connect an external speaker or microphone to a Micro:bit. These are required for the Micro:bit V1 since the Micro:bit V2 already has an inbuilt buzzer. However, even for V2, a user may connect an external speaker or headphones for better sound output.

Melodies

Music is made of melodies. The Micro:bit library is designed with a rich collection of pre-loaded melodies. We can play a simple melody on the Micro:bit using the music module. The module contains 21 melodies. Some of the most popular ones are as follows:

```
music.DADADADUM
music.PRELUDE
music.RINGTONE
music.BIRTHDAY
music.WEDDING
music.BADDY
```

```
music.POWER_UP
music.POWER_DOWN
```

We can access the complete list using the `dir(music)` command. Connect your Micro:bit to your PC and run the following simple MicroPython code to play any of these melodies:

```
import music
music.play(music.BLUES)
```

We can play multiple melodies by pressing different buttons on the Micro:bit. In the following code, we are playing two different melodies by pressing buttons **A** and **B** on the Micro:bit. The following code will play the FUNK melody when button **A** is pressed and the DADADADUM melody when button **B** is pressed:

```
import music
from microbit import *
while True:
    if button_a.is_pressed():
        display.show("1")
        music.play(music.FUNK)
    elif button_b.is_pressed():
        display.show("2")
        music.play(music.DADADADUM)
    else:
        display.show("0")
```

Let us explore one more program, where we will store the melodies as a playlist and play any melody in the playlist by pressing buttons **A** and **B**. We have defined a function, `play_music`. It is accessed by pressing the B button. This function will play different melodies based on the value of the index, i. The value of the index variable, i, is increased by pressing **A**. If the index value becomes 10, then it is reset to 0 because we have decided to play only 10 different melodies. The user can decide on the number of melodies:

```
import music
from microbit import *
i = 0
def play_music(i):
    print(i)
    if i == 0:
        music.play(music.DADADADUM)
    elif i == 1:
```

```
        music.play(music.ENTERTAINER)
    elif i == 2:
        music.play(music.PRELUDE)
    elif i == 3:
        music.play(music.ODE)
    elif i == 4:
        music.play(music.NYAN)
    elif i == 5:
        music.play(music.RINGTONE)
    elif i == 6:
        music.play(music.FUNK)
    elif i == 7:
        music.play(music.BLUES)
    elif i == 8:
        music.play(music.BIRTHDAY)
    elif i == 9:
        music.play(music.WEDDING)

while True:
    if button_a.is_pressed():
        i = i+1
        sleep(1000)
    elif button_b.is_pressed():
        play_music(i)
    else:
        display.show(i)

    if i== 10:
        i = 0
```

In this program, we have defined a function called `play_music()`. The argument to this function is an integer, `i`. In this function, we used the `if-elif-else` statement to declare 10 different conditional statements. For 10 different values of `i`, 10 different melodies are defined in this program. Therefore, depending on the argument's value to the function, the receptive melody will be played. Then, we check buttons **A** and **B** on the Micro:bit. If button **A** is pressed, then we increase the value of `i`, and the next melody is played, while if button **B** is pressed, the previous melody is played.

In this section, we used the Micro:bit to play a simple melody.

Custom melodies

With that, we have a good command of the `music` module of MicroPython. The melodies that are used in this module are made from various musical notes. A musical note is a symbol that donates musical sounds. A note can also represent a pitch class. Any music melody is made with a correct combination of musical notes. In English music, the notes are A, B, C, D, E, F, and G (the correct order is C, D, E, F, G, A, and B). You can learn more about the chromatic scale at `https://en.wikipedia.org/wiki/Chromatic_scale` and musical notes at `https://en.wikipedia.org/wiki/Musical_note`.

The next important aspect of melodies is the octave. The **octave** is the span between two of the same notes but with double the frequency. It indicates how high or low the note should be played. Octaves are in the range of 0 to 8. The higher the value of the octave, the longer the duration of the note. For example, if the octave is set to 6, then it will last two times longer than octave 3.

In MicroPython, the function to represent notes is as follows:

```
NOTE[octave] [:duration]
```

Here, `NOTE` represents the name of the note (A to G), `octave` is a number (0 to 8), and `duration` is the amount of time for which it is played. For example, D5:4 will play the note D in the fifth octave for a duration of 4.

Let us understand the notes and their duration with the help of *Table 12.1*:

Note Type	Whole Note	Half Note	Quarter Note	Eighth Note	Sixteenth Note
Symbol					
Name	Semibreve	Minim	Crotchet	Quaver	Semiquaver
Duration (in MicroPython)	16	8	4	2	1

Table 12.1 – Notes and their durations for MicroPython

Now, let us compose our first melody. As we mentioned previously, the sequence of notes in English music is C, D, E, F, G, A, and B. We will play the same sequence three times at octave number 4. The code is as follows:

```
from microbit import *
import music
```

```
for x in range(3):
    music.play(["C4", "D4", "E4", "F4", "G4", "A4", "B4"])
```

When you run this code, you will hear the sequence of tones making a melody three times. This is a good way for us to start composing melodies.

Now, let us compose one of the most commonly played melodies – that is, *Happy Birthday*. Refer to https://www.letsplaykidsmusic.com/happy-birthday-easy-piano-music/ to get the notes.

Now, we will store these note values in a list named `tune`, and then play the list values:

```
import music
tune = ["G4", "G4", "A3", "G4", "C1", "B2", "G4", "G4", "A3",
"G4", "D2", "C1"]
music.play(tune)
```

The preceding code will play the first 12 notes of the *Happy Birthday* melody. We have only mentioned the note names and octave number in this sequence. We have not mentioned the durations. To perfectly play the melody, we must also mention the correct durations.

Now, let us compose a melody by mentioning the durations as well. We need the help of a music teacher to know the duration for every note to play a melody with perfection. *Frère Jacques* is one of the most popular French rhymes. Let us compose it using the same function:

```
from microbit import *
import music
for x in range(3):
    music.play(["C4:4", "D4", "E4", "C4", "E4:4", "F4",
"G4:8"])
```

In this program, we have used seven different notes. One interesting aspect of this code is that we have not mentioned the duration for every note. MicroPython remembers the last defined duration and will allocate the same to the next notes. The first note is defined as C4:4. This means that the Micro:bit will play note *C* at octave 4 for a duration of 4. Now, if we want to play the next notes at the same octave and duration, then we can just mention the name of the note; writing the octave and duration is optional.

The following code will generate the same melody as the previous code. In this code, we have only defined the first note with full specifications. The next notes until F have not been defined with octave values and durations. We have mentioned the new value of G4 with a duration of 8:

```
from microbit import *
import music
```

```
for x in range(3):
    music.play(["C4:4", "D", "E", "C", "E", "F", "G4:8"])
```

In this section, we discussed how to create custom melodies by knowing the concepts of notes. If you have the technical knowledge to generate music through different notes, then you can apply that to generate electronic music using the Micro:bit.

Tempo and pitch

The next two features of music are tempo and pitch. **Tempo** is defined as the number of beats to be played per minute. Therefore, a piece of high-tempo music will play more beats per minute (bpm) and will sound fast. Similarly, if we keep the tempo value low, then we can hear the same melody as a piece of slow-paced music. In Micro:bit's MicroPython implementation, we have the `music.set_tempo(bpm=tempo)` function to set the tempo of the melody.

For example, the following code will play the melody at 100 **bpm**:

```
music.set_tempo(bpm=100)
```

Pitch is defined as the frequency of the tone. **Frequency** is the number of cycles in 1 second. We can't be confused about understanding tempo and pitch. Let us consider a note, A4, with a pitch of 440 Hz. This means that the speaker's membrane will vibrate 440 times every second to play this tone in octave 4. If we place the tone with A5, then the frequency or the pitch will double – that is, 880 Hz. Now, we can play this tone as a part of the melody at different tempo values.

Let us write a simple program to play a tone at different tempo levels. We will define a variable called `tempo` with a value of 90. Then, we will use the `music.set_tempo` function to set the tempo value equal to the variable value. Then, we will play the note C4 for a duration of 4 with a rest of 1 tick. We have used conditional statements with buttons A and B. We can increase the tempo with button A and decrease it with B. We can hear the same tone with a higher or lower tempo by pressing these two buttons:

```
from microbit import *
import music
tempo = 90
while True:
    music.set_tempo(bpm=tempo)
    music.play(['C4:4', 'r:1'])
    if button_a.was_pressed():
        tempo = tempo + 5
    if button_b.was_pressed():
        tempo = tempo -5
```

Now, let us explore another simple program where we can change the melody's tempo by using a potentiometer. The connections for the circuits are shown in *Figure 12.4*. We connect the extreme ends of the potentiometer to the ground and power supply (VCC) (3V, in this case). The central pin of the potentiometer is the output pin, which is connected to analog pin P0 of the Micro:bit:

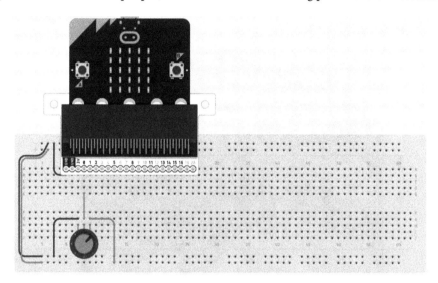

Figure 12.4 – The connections of a potentiometer to control the tempo of a melody

The program for the circuit is given here:

```
from microbit import *
import music
pot_val = 0
while True:
    pot_val = (pins.analog_read_pin(AnalogPin.P0) / 5)
    music.set_tempo(bpm=pot_val)
    music.play(['C4:4', 'r:1'])
```

Here, we defined a variable, pot_val, and set its value to 0. Then, we read the value of analog pin P0. The values read at this pin will be in the range of 0 to 1,023. The music tempo is generally in the range of 0 to 200 bpm. So, we have divided the analog pin value by 5 to map the read values from 0 to 200. Then, we set the tempo level to the value of pot_val and played the melody at this tempo level.

In this section, we learned about the tempo and pitch of the melodies. A piece of music can be played faster or slower, depending on the tempo values set for the same. This section helped us program the different tempo values to generate different melodies using MicroPython.

Working with a microphone

So far, we have generated different sound outputs using the Micro:bit. The Micro:bit is also equipped with a microphone. This microphone can be used to detect sound from the environment. This section will teach us how to handle speech using a Micro:bit. The microphone is present on the front of the Micro:bit V2. *Figure 12.5* shows its location:

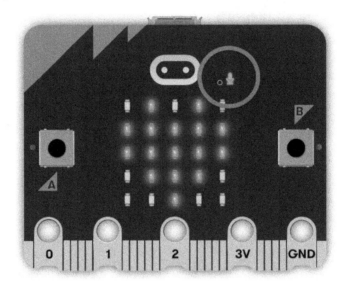

Figure 12.5 – Location of the microphone on the Micro:bit V2

(courtesy: https://microbit-micropython.readthedocs.io/en/v2-docs/_images/microphone.png)

The microphone that's built inside the Micro:bit has a sensitivity of -38dB ±3dB @ 94dB SPL. It has an SNR of 63dB. The microphone can detect frequencies from 100 Hz to 80 kHz. The complete datasheet for this microphone is available at https://www.knowles.com/docs/default-source/model-downloads/spu0410lr5h-qb-revh32421a731dff6ddbb37cff0000940c19.pdf?Status=Master&sfvrsn=cebd77b1_4.

In MicroPython, the microphone can respond to a predefined set of sound events. These events are based on the sound signal's wavelength and amplitude. All these events can be accessed with the SoundEvent class. We can use the microbit.SoundEvent function to access different events under this class. There are two main classes:

- microbit.SoundEvent.QUIET: This is used to detect the absence of sound or the transition from loud sounds to lower sounds or quietness.

- microbit.SoundEvent.LOUD: This function represents the presence of a loud sound around the microphone. It can also detect any short-duration sounds, such as claps.

Now, let us make a very simple system using the microphone. If we clap, the Micro:bit will detect it as a loud sound and display O on the LED matrix. If there is no sound detection, then X will be displayed:

```
from microbit import *
while True:
    if microphone.current_event() == SoundEvent.LOUD:
        display.show("O")
        sleep(1000)
    if microphone.current_event() == SoundEvent.QUIET:
        display.show("X")
```

Now, let us modify the program slightly and make a fancy disco light. The brightness of the LEDs can also be set using numbers from 0 to 255. In this program, we will detect the sound level using the `microphone.sound_level()` function. Then, we will display X on the LED matrix with varying intensities. For this purpose, we will declare an array named `imageX`, which will display X on the 5x5 LED matrix of the Micro:bit. We will multiply the matrix with the detected sound level; by doing so, the LEDs will glow with the computed intensities:

```
from microbit import *
imageX = Image("10001:"
               "01010:"
               "00100:"
               "01010:"
               "10001")
while True:
    display.show(imageX* microphone.sound_level())
```

Now, we will use another feature of the sound level that's been detected by the microphone. Assume that you are at your annual function and students are being called on stage and given awards. Some popular students receive applause for a longer duration, while others receive shorter ones. Using this system, we will count the duration of loud applause. The first decision is to set the loudness level to a certain value. As we mentioned previously, the microphone will detect sound levels from 0 to 255, so in this program, we have set the level to 200, which means only loud applause will be considered. Whenever a sound with a loudness of more than 200 is detected, then the `set_counter` variable will start recording the time from the `running_time()` function. This time will be recorded in milliseconds. Once the loudness has been reduced, we convert the time from milliseconds into seconds by dividing it by 1,000:

```
from microbit import *
microphone.set_threshold(SoundEvent.LOUD, 200)
```

```
set_timer = 0
while True:
    if microphone.was_event(SoundEvent.LOUD):
        set_timer = running_time()
        display.show(Image.HEART)


    if microphone.was_event(SoundEvent.QUIET):
        if set_timer > 0:
            time = running_time() - set_timer
            set_timer = 0
            display.clear()
            sleep(200)
            display.scroll(time / 1000)
```

In this section, we learned about the functionality of the in-built microphone of the Micro:bit. Then, we learned how to detect any sound using the microphone and the respective library and functions to detect the same. We also learned about the level of detected sounds and implemented interesting codes using those concepts. This section also taught us the related MicroPython concepts.

Working with speech

Earlier in this chapter, we generated music melodies using the Micro:bit. The Micro:bit can also be used to generate speech. This means we can generate words, sentences, and even poems using the Micro:bit. We will use a speech library to generate speech using the Micro:bit. Let us try generating a simple speech message using the speech library:

```
import speech
from microbit import *
speech.say("Hey!")
sleep(500)
speech.say("How are you friend")
sleep(1000)
```

This program will generate the speech *Hey! How are you friend?* We can hear the sound from the internal speaker or any external speaker connected to the Micro:bit.

The speech that's generated by the preceding program does so using the **Text-to-Speech** (**TTS**) default settings. The TTS conversion in the Micro:bit is done with **Software Automated Mouth** (**SAM**), which was originally released in 1982 for the Commodore 64. More details about SAM are available at https://simulationcorner.net/index.php?page=sam. This program can take up to 255 characters of textual input and generate a sound of around 2.5 seconds in one command.

The complete say function in the speech library can be accessed like so:

```
speech.say(words, *, pitch=64, speed=72, mouth=128, throat=128)
```

Here, words are the text input that the user provides. The preceding line of code mentions the default values of pitch, speed, mouth, and throat. If we want to change these values for better speech production, we must take care of the following parameters:

- The pitch parameter of the generated sound can be between 20 to 90 for practical speech generation. Anything below 20 is impractical to generate, and anything above 90 is a very low pitch. Therefore, we should keep it between 20 to 90. The best range is suggested to be 50 to 70, and 64 is the default pitch value.

- The speed parameter of the generated speech can be between 0 and 255. Any value below 20 is impractical, and anything above 100 is very slow. A speed of 70 to 75 is considered for normal conversation. We can reduce the value for faster speech production. The default value is 72.

- The mouth parameter of the generated speech represents how much the speaker is opening their mouth to generate the speech. A lower value represents the speaker speaking without moving their lips considerably. This number can be set between 0 and 255. The default value is kept in the center at 128.

- The throat parameter shows the stress on the speaker's throat while speaking. A lower value means the speaker is relaxed, while a higher number represents the speaker putting more effort into generating the speech. This number can be set between 0 and 255. The default value is kept in the center at 128.

The following code will generate the same sound, *Hello friend*, with five different settings:

```
import speech
from microbit import *
speech.say("Hello friend", pitch=60, speed=72, mouth=160,
throat=120)
speech.say("Hello friend", pitch=70, speed=92, mouth=190,
throat=120)
speech.say("Hello friend", pitch=80, speed=98, mouth=110,
throat=100)
speech.say("Hello friend", pitch=90, speed=64, mouth=120,
throat=100)
speech.say("Hello friend", pitch=98, speed=98, mouth=190,
throat=190)
```

In all these programs, we can see that the speech generated by the Micro:bit is inaccurate and not as close to human speech. We can use the pronounce() function from the speech library to generate a speech signal that's close to human speech. This function uses English phonemes to generate speech. A **phoneme** is the smallest unit of sound in speech. It can distinguish one word from another in a particular language. Every word is made up of multiple phonemes. For example, the word *hat* has three phonemes: *h*, *a*, and *t*. The English phonemes are coded with characters in the speech library. These codes are as follows:

SIMPLE VOWELS		VOICED CONSONANTS	
IY	f(ee)t	R	(r)ed
IH	p(i)n	L	a(ll)ow
EH	b(e)g	W	a(w)ay
AE	S(a)m	W	(wh)ale
AA	p(o)t	Y	(y)ou
AH	b(u)dget	M	Sa(m)
AO	t(al)k	N	ma(n)
OH	c(o)ne	NX	so(ng)
UH	b(oo)k	B	(b)ad
UX	l(oo)t	D	(d)og
ER	b(ir)d	G	a(g)ain
AX	gall(o)n	J	(j)u(dg)e
IX	dig(i)t	Z	(z)oo
		ZH	plea(s)ure
		V	se(v)en
DIPHTHONGS		DH	(th)en
EY	m(a)de		
AY	h(igh)		
OY	b(oy)		
AW	h(ow)	UNVOICED CONSONANTS	
OW	sl(ow)	S	(S)am
UW	cr(ew)	SH	fi(sh)
		F	(f)ish
		TH	(th)in
		P	(p)oke
SPECIAL PHONEMES		T	(t)alk
UL	sett(le) (=AXL)	K	(c)ake
UM	astron(om)y (=AXM)	CH	spee(ch)
UN	functi(on) (=AXN)	/H	a(h)ead
Q	kitt-en (glottal stop)		

Now, if we want to pronounce any word, we should combine the phonemes of that word and enter that as an argument to the pronounce() function.

Now, assume that we want to pronounce *hello* using phonemes. This word can be made using the combination of phonemes *H-EH-L-OH*, and our argument will be HEHLOH:

```
import speech
from microbit import *
speech.pronounce("/HEHLOH")
```

This articulation of *hello* is better than the previous one. Similarly, we can generate the phoneme codes for all the words by following the rules of phonetics and generating speech using the Micro:bit. You can refer to the text-to-phoneme table given at https://github.com/s-macke/SAM/wiki/ Text-to-phoneme-translation-table for more details.

Now, when we speak, sometimes, we elongate the pronunciation of a particular phoneme. For example, while calling someone standing close to us, we will say *Hey* for a short duration. Whereas if the person is far from us, we stretch our speech and say *Hey* for a longer duration. We can increase the duration of any phoneme by adding a multiplication factor to it. For example, *H-EH3* will generate the sound of *EH* for a three-times longer duration. The following code will generate the word *hello* in four different ways:

```
import speech
from microbit import *
speech.pronounce("/HEHLOH")
speech.pronounce("/HEH3LOH")
speech.pronounce("/HEH3LOH4")
speech.pronounce("/HEHLOH7")
```

In this section, we learned how to use the speech module of MicroPython and apply it to the Micro:bit. Then, we discussed the features of the speech library. We also discussed two different ways to produce speech using the Micro:bit. In the first approach, we used the say function with different settings. In the second approach, we worked with the concept of phonemes to generate the speech.

Summary

In this chapter, we explored a unique aspect of the Micro:bit: we learned how to use its inbuilt speaker to generate different sounds. We also talked about the connections of an external speaker and how they can be connected to the Micro:bit. Then, we learned how to generate a single musical tone with different octaves and durations. These tones can be combined to generate customized melodies. We also explored the `music` library of MicroPython to play some predefined melodies. In the second half of this chapter, we learned how to use a microphone and take sound inputs to perform interesting operations on the Micro:bit. Finally, we discussed how to generate human-like speech using the speech library.

In the next chapter, we will discuss networking and radio communication using the Micro:bit. We will explore in detail how two or more Micro:bit devices can communicate using the built-in radio. We will also create a few simple games using the radio communication capability of Micro:bit devices.

## 13

Networking and Radio

In the previous chapter, we learned how to program the Micro:bit for producing music and speech. We learned to connect an external speaker to the Micro:bit. We also explored the built-in speaker and microphone in Micro:bit V2.

Networking means connecting two or more computers (or microcontrollers). It enables them to exchange data. This short yet intense chapter focuses on networking multiple (a minimum of two) Micro:bit devices together to exchange data between them. We will also create a few simple games that demonstrate the radio communication capabilities of the Micro:bit. The following is a list of topics that we will explore in this chapter:

- A wired network with GPIO pins

- Working with radio

- Basketball using Micro:bit

- Rock, paper, scissors

- Voting using the Micro:bit

Let's get started with networking and radio communications.

Technical requirements

For this chapter, we will need a minimum of two Micro:bit devices with a computer connected to the internet and a micro USB cable. We will need the GPIO extender and jumper cables for the wired networking part.

A wired network with GPIO pins

We can use the GPIO pins of the Micro:bit to create very primitive networks that exchange data in binary format. We know that the *HIGH* signal (which corresponds to 1) and *LOW* signal (which corresponds to 0) represent binary data. We will use the built-in `write_digital()` method to exchange 1s and 0s between devices. Let's use two Micro:bit devices to create a simple circuit to exchange binary data:

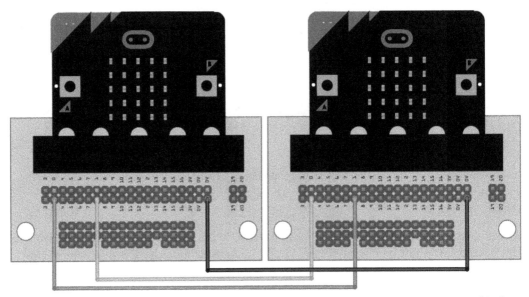

Figure 13.1 – Micro:bits connected together

Let's write a simple program that sends 1 over pin 0 if the **A** button is pressed and sends 0 otherwise. Similarly, we will read pin 1; if a device reads 1, then it will show a heart symbol and a smiley face otherwise. Here, pin 0 acts as a transmitter, and pin 1 acts as a receiver for both devices.

Upload this program to both devices and test it by pressing the **A** buttons on both devices:

```
from microbit import *
try:
    while True:
        if button_a.is_pressed():
            pin0.write_digital(1)
        else:
            pin0.write_digital(0)
        input = pin1.read_digital()
        if (input == 1):
            display.show(Image.HEART)
        else:
```

```
        display.show(Image.HAPPY)
except KeyboardInterrupt as e:
    print("Interrupted by the user...")
```

In the next section, we will explore radio communication with the Micro:bit in detail.

Working with radio

The Micro:bit comes with a built-in 2.4 GHz radio module. Using the built-in library radio, we can send and receive messages wirelessly. For demonstrations of the radio functionality, we do not need to create a circuit. We are ready to go by just powering up a few Micro:bit devices.

Turning the radio on and off

We can get started by turning the radio of a Micro:bit on and off with the following simple code executed in a REPL:

```
>>> from microbit import *
>>> import radio
>>> radio.on()
>>> radio.off()
```

Let's learn to send and receive messages.

Sending and receiving messages

We can send messages with the built-in radio.send() method. The maximum length of the string that can be sent with this method is 250 characters. Let's write the code for a transmitter routine:

```
from microbit import *
import radio
try:
    radio.on()
    while True:
        radio.send("Hello, World!")
        sleep(1000)
except KeyboardInterrupt as e:
    print("Interrupted by the user...")
```

Run this code on the first Micro:bit.

The built-in `radio.receive()` method receives messages at a Micro:bit receiver as shown in the following program:

```
from microbit import *
import radio
radio.on()
try:
    while True:
        msg = radio.receive()
        if msg is not None:
            display.show(msg)
        sleep(1000)
except KeyboardInterrupt as e:
    print("Interrupted by the user...")
```

Run this program on the second Micro:bit and see the radio communication in action.

We can change the program on the transmitter Micro:bit side to send different messages when a button is pressed as follows:

```
from microbit import *
import radio
try:
    radio.on()
    while True:
        if button_a.is_pressed():
            radio.send("A")
        elif button_b.is_pressed():
            radio.send("B")
        else:
            radio.send("C")
        sleep(1000)
except KeyboardInterrupt as e:
    print("Interrupted by the user...")
```

The receiver-side program requires no change. However, you may wish to add an `if-elif` ladder to decode the message on the receiver side. Then, you can write custom code for each type of message.

Using the knowledge we have gained so far, we can program a group of Micro:bit devices to behave like a group of fireflies. I have adapted the program written by Nicholas H. Tollervey

(https://microbit-micropython.readthedocs.io/en/v2-docs/tutorials/
radio.html) to work for a group of two Micro:bit devices. Run the following program on a
minimum of two Micro:bit devices:

```
import radio, random
from microbit import *

flash = [Image().invert()*(i/9) for i in range(9, -1, -1)]
print(flash)

try:
    radio.on()
    while True:
        if button_a.was_pressed():
            radio.send('flash')
            print("Sent the flash message!")
        if radio.receive() == 'flash':
            sleep(random.randint(50, 1000))
            display.show(flash, delay=100, wait=False)
            sleep(random.randint(50, 1000))
            radio.send('flash')
except KeyboardInterrupt as e:
    print("Interrupted by the user..")
```

Once the program runs on the devices, if you press the **A** button on any Micro:bit running this
program, all the devices will start flashing in a continuous loop.

In this section, we talked about the basic GPIO pin-based network connections. We learned how to set
up radio communication on a Micro:bit and use it for simple applications. We are able to communicate
between two Micro:bits now using radio communication. This section gave us the confidence to build
radio communication-based games and applications and we will use this knowledge in future sections.

Basketball using Micro:bit

In this activity, we will make a fun game using radio communication between two Micro:bits. Both
players will hold a Micro:bit in their hands. One of the users will have a ball shape displayed on their
Micro:bit. If the user shakes their Micro:bit, the message will be sent to the other Micro:bit, and the
display will show a ball on the other user's Micro:bit. The display of the first user will go blank. It shows
that the first user has transferred the ball to the second user. The second user can also shake their

Micro:bit to pass the ball back to the first user. *Figure 13.2* shows the arrangement for the basketball passing game:

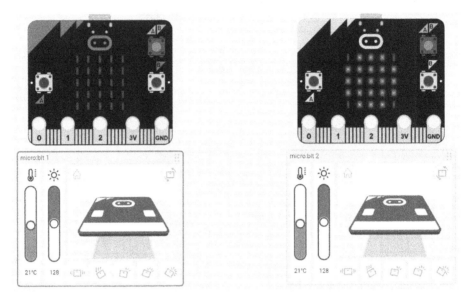

Figure 13.2 – Passing the basketball from one Micro:bit to another

For coding, we will code both the Micro:bits with the same code. We define a custom display using the Image () function. The function will define a 5 x 5 image. The intensity of the LED can be set between 0 to 8. Here, a 0 value will keep the LED off, and the LED will glow at maximum brightness if set to 8. Therefore, we have defined a ball shape with the help of the following array named ball:

```
("08880:"
 "88888:"
 "88888:"
 "88888:"
 "08880")
```

Now, we display the ball shape at the beginning. When the Micro:bit is shaken, then we send the ball code on the radio and clear the display. Whenever we receive the ball code, we display the shape again:

```
from microbit import *
import radio
ball = Image("08880:"
             "88888:"
```

```
            "88888:"
            "88888:"
            "08880")
radio.on()
display.show(ball)
while True:
    if accelerometer.was_gesture('shake'):
        radio.send('ball')
        display.show(" ")
    incoming = radio.receive()
    if incoming == 'ball':
        display.show(ball)
```

In this section, we made a simple game that involves radio communication between two or multiple Micro:bits. In this example, we displayed a ball shape on the LED matrix and after shaking it, the shape disappeared from the Micro:bit and sent a signal to another Micro:bit. This signal was received at the next Micro:bit and the ball shape was displayed on the LED matrix. It appears as though we have passed the ball from one Micro:bit to another.

Rock, paper, scissors

Rock, paper, scissors is a very popular game; almost all of us play it with friends. We can make three gestures with our hands – that is, rock, paper, and scissors. Every gesture can overpower one of the other two gestures and can be defeated by another. Two users will display a random gesture, and the following rules decide the winner:

- Rock can kill scissors but is wrapped in paper
- Paper is cut by scissors but can cover rock
- Scissors can cut paper but are broken by rock
- If both gestures are the same, then it is a tie

For coding, we will assign integers to these three gestures and use conditional statements to decide the winners.

We have used the following codes and image icons for the gestures:

- Rock = 0 and displays a square
- Paper = 1 and displays a heart
- Scissors = 2 and displays scissors

We will also display three different outputs on the screen based on the game's outcome. We select the pre-defined gestures from the Image library. The following display images are chosen for the respective outcomes:

- Asleep = match is tied

- Tick symbol = user won

- Cross symbol = user lost

Figure 13.3 shows one such game between user 1 and user 2:

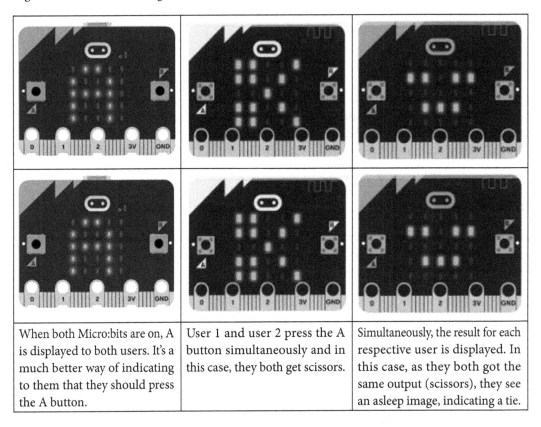

When both Micro:bits are on, A is displayed to both users. It's a much better way of indicating to them that they should press the A button.	User 1 and user 2 press the A button simultaneously and in this case, they both get scissors.	Simultaneously, the result for each respective user is displayed. In this case, as they both got the same output (scissors), they see an asleep image, indicating a tie.

Figure 13.3 – A match tied between user 1 and user 2

Initially, user 1 selected a random gesture (the scissors icon). Now, user 1 will confirm the selection. After this, user 2 will select a random gesture (again, the scissors icon) and confirm it. The figure shows that the match was a tie in this case.

Figure 13.4 shows one more game between user 1 and user 2:

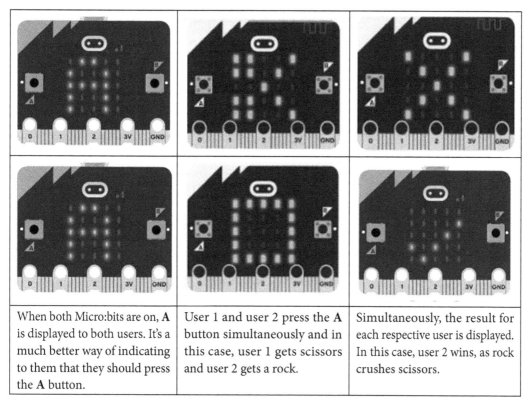

When both Micro:bits are on, **A** is displayed to both users. It's a much better way of indicating to them that they should press the **A** button.	User 1 and user 2 press the **A** button simultaneously and in this case, user 1 gets scissors and user 2 gets a rock.	Simultaneously, the result for each respective user is displayed. In this case, user 2 wins, as rock crushes scissors.

Figure 13.4 – State of the Micro:bit at various junctures in the game (user 1 and user 2 – user 2 WINS)

Initially, user 1 selected a random gesture (rock with a square icon). Now, user 1 will confirm the selection. After this, user 2 will select a random gesture (paper with a heart icon) and confirm it. The preceding figure shows that user 1 has lost the match.

Now, let us consider the coding aspects. As decided earlier, we define the icons for rock, paper, and scissors, respectively. We define a my_move variable. When the **A** button is pressed, the variable will store a random number (0, 1, or 2). The user will see the respective icon on their screen. Then, after a second, the number is sent to the other board using the radio.send function.

Once the message is received, the message is stored in the opp_move variable. Once we have both values (my_move and opp_move), then we compare them and display the outcome on the Micro:bit: The code here should be uploaded to both the Micro:bits. Initially, we called the required libraries and defined the scissor, rock, and paper display patterns:

```
from microbit import *
import random
```

```
import radio

my_move = 0
opp_move = 0
SCISSORS = Image("88008:"
                 "88080:"
                 "88800:"
                 "88080:"
                 "88008")
radio.on()

display.show("A")  #on start display letter A
```

Now, we define the function by pressing the **A** button. We generate a random number between 0 to 2 and depending on the number, one of the moves is selected as my_move:

```
while True:
  if button_a.was_pressed():
    my_move = random.randint(0, 2)
    if my_move == 0:
      display.show(Image.SQUARE)
    elif my_move == 1:
      display.show(Image.HEART)
    else:
      display.show(SCISSORS)
    sleep(1000)
    display.show(
      " "
    )
    radio.send(str(my_move))
```

Now, we have sent the value of my_move via radio communication.

The following code explains the logic followed on receiving the opponent's move. We will compare the value of my_move with opp_move and set all nine possibilities:

```
opp_move = radio.receive()

if my_move == 0:
```

```
    if opp_move == "0":
       display.show(Image.ASLEEP)
    elif opp_move == "1":
       display.show(Image.NO)
    elif opp_move == "2":
display.show(Image.YES)
#Now, we check if the input is 1
  elif my_move == 1:
    if opp_move == "0":
       display.show(Image.YES)
    elif opp_move == "1":
       display.show(Image.ASLEEP)
    elif opp_move == "2
       display.show(Image.NO)
#And here is the last case, when input is 2
  else:
    if opp_move == "2":
       display.show(Image.ASLEEP)
    elif opp_move == "1":
       display.show(Image.YES)
    elif opp_move == "0":
       display.show(Image.NO)
```

In this section, we designed a very popular game using Micro:bits. We replicated the rock, paper, scissors game using a combination of two Micro:bits. We used radio communication to transfer a random value for rock, paper, or scissors from one Micro:bit to another. By comparing the user's move and the opponent's move, the result is displayed on the LED matrix.

Voting using the Micro:bit

Now, let us make an application involving more than two Micro:bits. We can create an electronic voting machine using multiple Micro:bits. One Micro:bit is used to collect the votes and others are used to cast their votes. Only one Micro:bit will be used to receive the radio messages, while all others will be used to send the radio message. The codes for the two categories of Micro:bits will be different. The code for sending the vote from a Micro:bit is as follows:

```
from microbit import *
import radio
radio.on()
```

```
my_vote = "Vote"
while True:
    if button_a.was_pressed():
        my_vote = "Yes"
        display.show(Image.YES)
        radio.send(my_vote)
    elif button_b.was_pressed():
        my_vote = "No"
        display.show(Image.NO)
        radio.send(my_vote)
```

In this code, we have chosen two buttons, **A** and **B**, to decide the value of the my_vote variable. We send Yes by pressing the **A** button and No with the **B** button. Now, let us explore the code for receiving and counting the casted votes:

```
from microbit import *
import radio
radio.on()
my_vote = "Vote"
vote_count = 0
while True:
    my_vote = radio.receive()
    if my_vote == "Yes":
        vote_count = vote_count+1
        display.show(vote_count)
    elif my_vote == "No":
        vote_count = vote_count
        display.show(vote_count)
```

In this section, we designed another simple yet useful game. This game is perfect for digital voting in a small group. Every user is given a Micro:bit. There is one Micro:bit that counts the votes while others are used to send their votes.

Summary

This chapter explored the basics of wired networking and radio communication with Micro:bits `radio` library. Once we understand how to work with basic communication, we can use this concept with the combination of earlier hardware projects we learned. We can use this concept in plenty of projects involving a group of Micro:bit devices. For example, we can use it with a few external inputs (push buttons and analog inputs) and external outputs (LEDs, RGB LEDs, NeoPixels, and Piezzo buzzers) to create interesting projects.

The next chapter explores different sensors built into the Micro:bit. We will discuss the access and use of capacitive touch and a temperature sensor or a light sensor on the Micro:bit. We will learn how to locate, access, and use these sensors and develop interesting applications using them.

Further reading

We can find more information about the radio API in a Micro:bit implementation of MicroPython at `https://microbit-micropython.readthedocs.io/en/v2-docs/tutorials/radio.html` and `https://microbit-micropython.readthedocs.io/en/v2-docs/radio.html`.

14

Advanced Features of the Micro:bit

We have explored a variety of applications of using the Micro:bit throughout this book. However, we have not explored the complete set of features available on the Micro:bit board. A variety of sensors are built into the board that can be used to perform interesting operations. In other microcontrollers, all such sensors are not inbuilt, so we must attach them separately. External sensor integration consumes a greater number of pins as well as power. Therefore, the Micro:bit has a huge advantage over other similar boards.

In this chapter, we will learn about the sensors built into the Micro:bit. Some of these features are present in both versions, while a few are only present in V2. We will learn how to access each sensor and control some output devices connected at different pins. We can also display useful information from these sensors on the LED matrix. We will discuss the following sensors of the Micro:bit:

- Capacitive touch
- Temperature sensor
- Light sensor

Let us get started by looking at the features of the Micro:bit with MicroPython.

Technical requirements

Apart from the usual setup, the demonstrations in this chapter need the following components:

- External batteries
- A torch for providing artificial light
- A flame or heat source for temperature control

Capacitive touch

Touch is one of our basic senses. We touch various objects with our fingers to feel them and obtain information about their texture, temperature, and other surface properties. On the Micro:bit, round pins 0, 1, and 2 can be touched to provide input from the human touch. These pins are available on both versions of the Micro:bit. They work on the principle of resistive touch. When we touch these pins, a connection to the ground of the Earth is established through our fingers. Therefore, the ground is always required while providing inputs from these pins.

In this chapter, we will learn about the touch sensor, which is only supported on V2. The sensor is present on the front side of the Micro:bit. Its location is above the LED matrix, as shown in *Figure 14.1*:

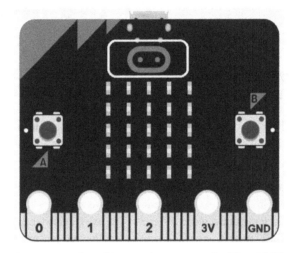

Figure 14.1 – Capacitive touch sensor on the Micro:bit V2

Any capacitive surface activates the capacitive touch sensor. Our skin is a capacitive surface as we access modern smartphones through the same capacitive touch technology. Capacitive touch does not require a ground connection. Let us make a very simple application using the Micro:bit capacitive touch sensor. We will display the heart image if we touch the sensor and a blank display if we remove our finger from the sensor. We will use the `pin_logo` module to detect and access the capacitive touch on the Micro:bit V2.

In this code, initially, we must explicitly declare that `pin_logo` will accept capacitive touch by using the following command:

```
pin_logo.set_touch_mode(pin_logo.CAPACITIVE)
```

Then, the code is a simple conditional statement that is triggered by the detection of touch on `pin_logo`. If the touch is not detected, the display will turn blank:

```
from microbit import *
pin_logo.set_touch_mode(pin_logo.CAPACITIVE)

while True:
    if pin_logo.is_touched():
        display.show(Image.HEART)
    else:
        display.clear()
```

Now, let us make a slight modification to the code. We will set a timer to count the amount of time for which the touch logo was touched. We will declare a variable named `set_timer` to count the running time. When it is no longer being touched, the time will be displayed in seconds:

```
from microbit import *
pin_logo.set_touch_mode(pin_logo.CAPACITIVE)
set_timer = 0
while True:
    if pin_logo.is_touched():
        display.show(Image.HEART)
        sleep(50)
        set_timer = running_time()
    else:
        if set_timer > 0:
            time = running_time() - set_timer
            set_timer = 0
            display.clear()
            sleep(200)
            display.scroll(time / 1000)
```

In this section, we talked about the capacitive touch on the Micro:bit V2. This logo can be activated by the touch of human skin. This logo can be used to design any system that can be controlled by the touch response. We also designed a very simple application, where we displayed LED patterns upon touching the logo and displayed the amount of time for which the logo was touched.

Temperature sensor

A **temperature sensor** is a sensor that measures the temperature of any surface or its surroundings. It is an input device that provides information related to the temperature. We can find temperature sensors around us in a variety of appliances. For example, air conditioning works purely on the value of the surrounding temperature. The temperature sensors inside the appliance measure this temperature, and based on the temperature, the air conditioner's compressor is turned on or off. The temperature sensor in industrial applications is also known as a **thermostat**. Most of the temperature sensors are integrated with a display to show the sensor's output (measure temperature).

The BBC Micro:bit has a temperature sensor inbuilt inside the main processor. The main processor is located on the back of the Micro:bit. *Figure 14.2* shows the processor's location on a Micro:bit:

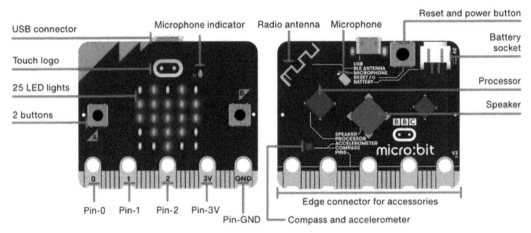

Figure 14.2 – Location of the main processor on a Micro:bit (courtesy:
https://microbit.org/get-started/user-guide/overview/)

This sensor displays the internal temperature of the device in degrees Celsius. This temperature is also a good estimate of the temperature around the device. So long as the processor is not overloaded with heavy computational load, the temperature of the processor is almost the same as the surrounding temperature. So, we can program our Micro:bit so that it can sense the temperature, display that information on the LED matrix or any external display, and even trigger some actuators based on the temperature value.

Let us write some simple code to read the temperature value and display it on the LED matrix of the Micro:bit. We do not need any additional libraries to access the temperature sensor of the Micro:bit. In this code, we have declared a new variable named Temp:

```
from microbit import *
Temp = temperature()
while True:
```

```
display.show('.')
Temp = temperature()
if button_a.was_pressed():
    display.scroll(Temp)
sleep(1000)
display.clear()
```

Now, let us make another application using the temperature sensor. We will continuously monitor the temperature and store it in the Temp variable. We will declare two new variables, namely MAX and MIN. These two variables have been created to store the minimum and maximum temperature observed during the whole monitoring time. We can take the Micro:bit to different places to observe these changes. Logically, we assign the value of the observed temperature (the Temp variable) to both variables at the beginning. Whenever the observed temperature is below the value of MIN, then we update the value of MIN to this newly observed value.

Similarly, if the value of the observed temperature is more than MAX at any time, then we must update the value of MAX to the newly observed value. We can display the maximum temperature by pressing button **A** and the minimum temperature by pressing **B**. The code for the system is as follows:

```
from microbit import *
Temp = temperature()
MAX = Temp
MIN = Temp
while True:
    display.show('.')
    Temp = temperature()
    if Temp < min:
        min = Temp
    if Temp > max:
        max = Temp
    if button_a.was_pressed():
        display.scroll(MAX)
    if button_b.was_pressed():
        display.scroll(MIN)
    sleep(1000)
    display.clear()
```

Figure 14.3 shows the possible simulations related to this system. In this case, the initial temperature was 25 degrees Celsius. We have varied the temperature between 12 to 35 degrees. Therefore, by pressing the related buttons, we can also display the minimum and maximum temperatures:

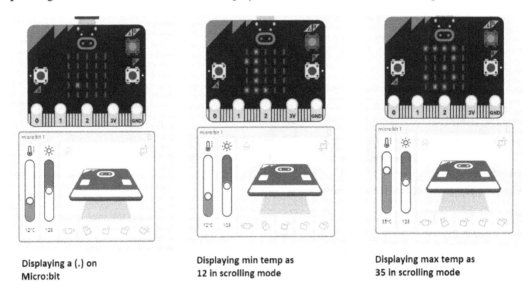

Displaying a (.) on
Micro:bit

Displaying min temp as
12 in scrolling mode

Displaying max temp as
35 in scrolling mode

Figure 14.3 – Displaying the minimum and maximum temperature of the BBC Micro:bit

Now, we have a good idea about reading the temperature values from the Micro:bit. Let us integrate some actuators into the Micro:bit and control them with the temperature values. In this activity, we will design a simple temperature-controlled fan. If the temperature exceeds a certain threshold, we will turn on the fan to cool the surroundings. As the fan starts operating, the temperature will start coming down. And, if the temperature falls below the set threshold, the fan will be turned off. In the following code, we have set the threshold temperature at 27 degrees Celsius. The Micro:bit will display H if the temperature is more than 27 degrees and turn on the fan using analog pin 1. If the temperature is less than 27 degrees, then we will display L and turn off the fan:

```
from microbit import *
Temp = temperature()
while True:
    Temp = temperature()
    if Temp > 27:
        display.show("H")
        pin1.write_analog(255)
        sleep(1000)
    else:
```

```
        display.show("L")
        pin1.write_digital(0)
        sleep(1000)
    display.clear()
```

In this section, we discussed the temperature sensor present inside the Micro:bit. This sensor is present in both versions. In V2, it is located on the microprocessor. The temperature sensor provides the temperature in degrees Celsius. Using the MicroPython library, we can read the temperature and store it in a variable. We have designed simple applications using the temperature recorded from the Micro:bit, displaying different patterns based on the recorded temperature.

Light sensor

The Micro:bit V2 is a sensor-rich board. We know that there are 25 LEDs on the front to display anything using a combination of these LEDs. These 25 LEDs also have built-in light sensors to detect the intensity of light falling on them. The intensity of the light falling on these LEDs is read between 0 to 255. This means the highest intensity value will be read as 255, and the complete darkness will be read as 0. All the intensities in between are equally spaced in between these values.

Figure 14.4 shows the location of the light sensors on a Micro:bit:

Figure 14.4 – Position of light sensors on a Micro:bit (courtesy: https://
microbit.org/projects/make-it-code-it/sunlight-sensor/)

> **Important note**
> No fixed formula relates the intensity of light with the value. It will completely depend on our observations and adjustments for it to work properly with the light intensity values.

Let us write some simple code that will read the light level on the surface of the Micro:bit and display it on the Micro:bit. Since the number can be between 0 and 255, we can use the `scroll` function to display the value:

```
from microbit import *
while True:
    lightLevel = display.read_light_level()
    display.scroll(lightLevel)
    sleep(2000)
```

Now, we can read the value of the light intensity. Let us improve our design and develop a simple application using the same concept. In smart homes, there's a very popular device called an automatic lighting system. The system detects the intensity of light in the surroundings, and if the intensity falls below a certain threshold, then the lights are switched on in that area. Such applications are popularly used for outdoor lights. For example, whenever the light intensity falls below a certain value in the evening, the outdoor lights will be switched on. And, as the light intensity increases, the lights will switch off automatically in the morning. This simple system saves a lot of electricity.

In this program, we will define a matrix using the `Image` function, where all the values are set to 8 to define the maximum output of the LEDs. Whenever light intensity is below a set threshold, all 25 LEDs are turned on. The intensity is decided based on the observed intensity values at the system's location. Here, we have set the value to `100`:

```
from microbit import *
while True:
    if display.read_light_level() < 100:
        display.show(Image(
        "88888:"
        "88888:"
        "88888:"
        "88888:"
        "88888"))
    else:
        display.clear()
    sleep(5000)
```

We can also attach any external lights via relays and switch them on using the same logic. *Figure 14.5* shows an external LED connected to the Micro:bit. The external LED represents the streetlight (outdoor light) of the system:

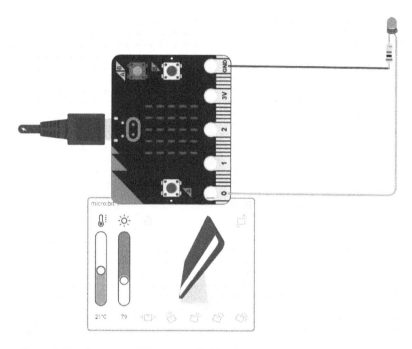

Figure 14.5 – An external LED controlled by the intensity of surrounding lights

The program for the automated streetlight is simple and is shown here. We select a light intensity threshold value and set the digital pin 0 to high or low, depending on this value. The digital pin is set to **HIGH** for lower values; for higher values, it is set to **LOW**:

```
from microbit import *
while True:
    if display.read_light_level() < 100:
        display.show("L")
        pin1.write_digital(1)
        sleep(1000)
    else:
        display.show("H")
        pin1.write_digital(0)
        sleep(1000)
```

The light sensor can also be combined with the inbuilt speaker of the Micro:bit to develop a customized morning alarm. We will press button **A** at night to set the alarm. Whenever sunrise is detected (the indicator is that the light intensity goes high), the speaker can play a melody or even speak a customized message. In *Chapter 12, Producing Music and Speech*, we learned how to generate melodies and speech. We can combine these concepts to make a custom alarm for ourselves. The code for the alarm is as follows:

```
from microbit import *
import music
status = 0
while True:
    if (display.read_light_level()<100):
        display.show("N")
        sleep(100)
        if (button_a.is_pressed()):
            status = 1
            while status == 1:
                display.show("L")
                music.play(music.RINGTONE)
                if (button_b.is_pressed()):
                    display.show("Good Morning")
                    status = 0
                    sleep(1000)
```

The Micro:bit has an inbuilt light sensor on the front of the board. In this section, we discussed how to use the light sensor and read the light intensity falling on the board. Based on this light intensity, we designed a simple daylight alarm. This alarm can also be integrated with the onboard speaker and generate a musical tone when the alarm is triggered.

Summary

In this concise chapter, we explored a few additional unique features of the Micro:bit. First, we learned how to use the capacitive touch-enabled logo on the front of the Micro:bit. This capacitive touch logo does not require a ground connection. We also discussed the inbuilt temperature sensor. The sensor is inside the processor and gives a very good estimate of the surrounding temperature in degrees Celsius. We built applications using the observed temperature. In the end, we discussed the use of inbuilt light sensors. These sensors are present along with the LED matrix on the front of the Micro:bit. We developed an interesting alarm using this concept.

The next chapter will explore sewable components and wearable computing with the Micro:bit and MicroPython. We will also have a brief look at more software programming platforms that use the Micro:bit.

15

Wearable Computing and More Programming Environments

In the previous chapter, we explored and demonstrated the hardware features of the Micro:bit. We also demonstrated the functionality of many onboard sensors. Now, we will develop some new applications using the Micro:bit where we will carry the Micro:bit with us. Initially, we will discuss how to make a simple pedometer using a Micro:bit. We will also learn how to detect a falling movement using similar concepts. Then, we explore an interesting topic known as **sewable computing**. This includes attaching or sewing computing devices to our clothes and carrying them with us.

The following list of topics will be explored in this chapter:

- Programming a pedometer using a Micro:bit
- Fall detector
- Sewable and wearable computing
- More programming frameworks

Let us get started with wearable computing.

Technical requirements

For this chapter, we will need a Micro:bit device with the following components:

- Conductive thread
- An embroidery hoop and a cotton cloth
- Sewable electrical components such as LEDs, switches, push buttons, and so on

Programming a pedometer using a Micro:bit

A **pedometer** is a device that measures the number of steps taken by a person while walking or running. These devices are in high demand in the modern world as many people monitor their daily physical activities with electronic devices. These observations help them manage their calorie intake. In a Micro:bit, we have an accelerometer to monitor various physical motions.

The concept of measuring the number of steps taken by the user is based on the physical motion detected by the Micro:bit. We can attach the Micro:bit to the foot or shoe of the user. A battery pack is attached to the Micro:bit. The Micro:bit will detect the *shake* movement whenever a person takes their next step. For every such movement that's detected, we can increase the count of the number of steps taken. Their current step count is displayed on the LED matrix. We can reset the count to 0 by pressing button **A**. *Figure 15.1* shows a Micro:bit attached to a shoe, which specifies the current count of steps:

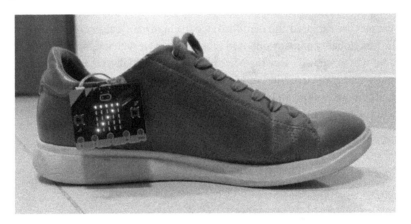

Figure 15.1 – A Micro:bit mounted on a shoe to count the number of steps the user takes

The code for this program is as follows:

```
from microbit import *
steps_count=0
while True:
    if button_a.is_pressed():
        steps_count =0
        display.show(steps_count)
    if accelerometer.was_gesture('shake'):
        steps_count += 1
        display.show(steps_count)
```

Sometimes, we can also get shaky movements due to non-walking activities, such as standing, a leg being shaken, or similar activities. In that case, the value of `steps_count` can increase even without taking a step. We can also design this pedometer with the help of the y axis data of the accelerometer. If the Micro:bit is attached to the shoe in such a way that the vertical movements of the foot cause the y axis value to change, then we will only consider a step has been taken if the change in the y axis value is more than a certain threshold. We will consider a change of more than `1000` in the y axis as one step that's been taken. This change can be observed when a person takes one stride. Let us have a look at the code now:

```
from microbit import *
steps_count=0
while True:
    if button_a.is_pressed():
        steps_count =0
        display.show(steps_count)
    if accelerometer.get_y() > 1000:
        steps_count += 1
        display.show(steps_count)
```

In this section, we talked about a very simple application of wearable computing: a pedometer. This system can measure the distance traveled by the user. We designed a pedometer to calculate the number of steps the user takes. This design was based on the accelerometer, which is built into the Micro:bit. The vertical stride of a step will bring a change in the y axis value. Based on this change, we count the number of steps.

Fall detector

Now, let us design another helpful device using the Micro:bit. Elderly people, children, and people with balancing issues are prone to falling while doing their daily activities. Such people require extra attention or a permanent attendee to monitor their movement. This person should monitor their movements and help them in case they fall. Due to a variety of reasons, the presence of an additional person might not be possible 24/7. We can design a fall detector using the Micro:bit that will generate a signal for the other person if the user has fallen. This device can also be used to detect accidents on bicycles or any other fall. In such cases, the device can call for help.

For this experiment, we need to check the range of accelerometer data from the Micro:bit for different movements. We will use the Mu editor and write a simple program to print and plot the accelerometer values for analysis purposes:

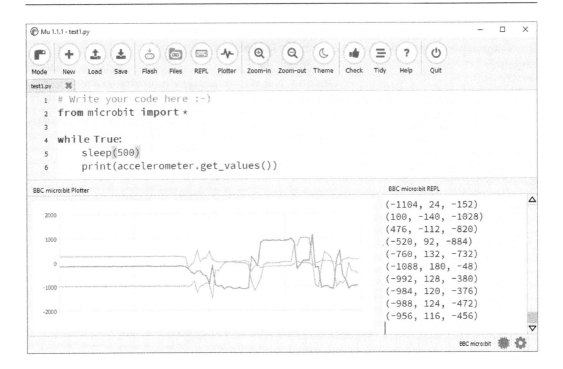

Figure 15.2 – Accelerometer data from the Micro:bit plotted using the Mu editor

Figure 15.2 shows the numerical and graphical representations of the x, *y*, and *z* axes for the accelerometer data. In the Mu editor, we will write the code, then click the **Flash** button to transfer the code to the Micro:bit. Then, we will click the **REPL** and **Plotter** buttons to visualize the accelerometer data in the form of numbers and graphs.

The code to generate this data is as follows:

```
from microbit import *
while True:
    sleep(500)
    print(accelerometer.get_values())
```

Now, let us develop the fall detector system using this concept. We should connect our Micro:bit to the body of the person. Now, when the user falls, which axis values change the most? This will depend on the orientation of the Micro:bit when it is attached to the user's body. To detect this fall, we should generate an alarm sound. Assuming that the *z* axis changes the most, we can write the following code:

```
from microbit import *
import music
```

```
while True:
    if accelerometer.get_z() > 800:
        music.play(music.DADADADUM)
        sleep(100)
```

In this section, we designed a helpful system for elderly people. The Micro:bit is attached to the body of the user in such a way that if they fall, the accelerometer's *z* axis value will change. Depending on the detection of the fall, the alarm can be raised for others.

Sewable and wearable computing

There is a whole world of sewable and wearable projects out there. Wearable projects can be worn on the body (in the form of lighting and sensors; it is so cool to see them in action!). To make this happen, we have to use sewable components. Sewable components can be sewn into fabrics using special materials. That is why they are known as **sewable components**. In this section, we will provide a brief overview of sewable components and prototyping.

Let us start with the base. We can use any fabric made of non-conductive material such as cotton, felt, or worsted wool. We can even use an embroidery hoop, as follows:

Figure 15.3 – Embroidery hoop with art in progress (courtesy: https://www.flickr.com/photos/
hey__paul/8355465843/ by Hey Paul Studios at https://www.flickr.com/people/45257015@N03,
made available under the CC BY license at https://creativecommons.org/licenses/by/2.0/deed.en)

Another essential thing we need is a **conductive thread**. As its name suggests, it's a sewing thread made of conductive and flexible material. It carries power and signal to the components sewn on the non-conductive fabric. The following figure shows a bundle of conductive thread:

Figure 15.4 – Conductive thread (courtesy: https://upload.wikimedia.
org/wikipedia/commons/e/ee/Conductive_thread.jpg)

We also have an assortment of electrical components, such as LEDs, RGB LEDs, WS2812, push buttons, and switches in sewable form, as shown here:

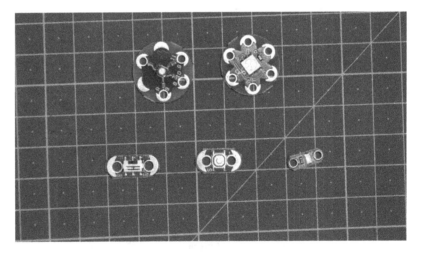

Figure 15.5 – Sewable components

BBC's Micro:bit may not be the best platform for sewable and wearable computing. However, we can use it for wearable projects with a little effort and ingenuity. The following figure shows a wearable project in progress:

Figure 15.6 – Wearable project using an Arduino Lilypad (courtesy: https://upload.
wikimedia.org/wikipedia/commons/e/ea/Lilypad_Arduino_with_fading_LEDs.jpg)

In *Figure 15.6*, we can see the Arduino Lilypad (a retired product) and LEDs in action. We can use the 0, 1, 2, GND, and 3V pins of the Micro:bit for the sewable project directly. Here, we have to power the Micro:bit with a portable battery pack or a coin cell. Also, as mentioned earlier, with a little bit of ingenuity, we can use the rest of the pins of the Micro:bit to suit our purpose. We can tape the conductive thread with a jumper cable that has a female header, as shown in the following figure:

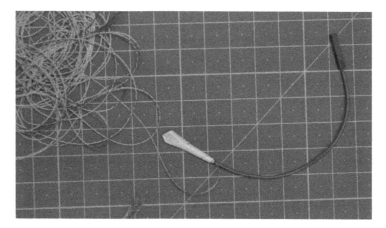

Figure 15.7 – Modified jumper

We can use these modified jumper cables with the combination of Micro:bit and edge connector to use all the possible pins of Micro:bit for our creative project. The conductive thread part of such modified cables will be used for sewing, and the female jumper header part will be used to connect the circuit of our creative sewing project with the edge connector.

Before you start sewing with the conductive thread, you may wish to build a prototype of the circuit using crocodile clips as shown in the following figure:

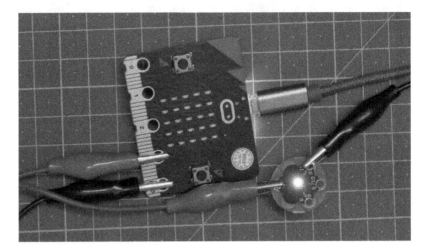

Figure 15.8 – Prototyping with crocodile clips

Figure 15.8 shows the green-colored section of an RGB LED (in sewable form) powered by the 3V and GND pins of the Micro:bit.

> **Important note**
> Use crocodile clips to build prototypes, not for the project itself; the clips don't work very well in wearable projects. For actual wearable projects, use conductive thread.

Companies such as Adafruit and SparkFun produce all the components that are required for wearable computing. You can get clones of these components from online marketplaces.

With that, we've learned how to use a Micro:bit with sewable components to create prototypes and actual wearable projects. As an exercise for this chapter, try building a few simple wearable projects such as a purse or a jacket, and add the RGB LEDs powered by the Micro:bit to add life to the projects.

More programming frameworks

This book is dedicated to learning MicroPython. However, MicroPython is not the only supported programming platform for programming a Micro:bit. We can use many other platforms. The following is a list of programming frameworks that support the Micro:bit:

- Microsoft MakeCode
- Scratch

- C++
- JavaScript

Now, let's summarize this chapter.

Summary

In this chapter, we explored a unique feature of the Micro:bit. Initially, we explored how can we use the Micro:bit's accelerometer and develop some real-world applications such as a step counter or a program that detects if a person has fallen. Then, we learned about sewable computing. Sewable computing includes attaching computing nodes and sensors to our clothes with the help of electrically conductive threads so that we can wear them all the time and the relevant data is transmitted to other devices. These devices are very useful for health monitoring and related purposes.

Conclusion

We are pleased that you have reached the end of this book. This book taught you about a very vibrant and compact computing board called the Micro:bit. The BBC develops the board, and it is widely used across the globe for education and development purposes. In this book, we learned how to use this board by implementing its input-output pins, LED buttons, and various sensors. This board is feature rich. A variety of sensors are already available on the board, which reduces the need to depend on other peripheral devices for simple computing purposes. We also learned how to attach external displays. A Micro:bit is capable of listening to our voice commands and even generating speech signals, and we developed interesting applications by implementing these components. We hope that the learning experience in this book has opened the doors for your imagination to be converted into real projects. This book is just the start, and we are sure that you will be developing even more interesting applications using the Micro:bit board.

Further reading

An extensive list of the programming frameworks supported by the Micro:bit and links to the necessary resources has been compiled by Carlos Pereira Atencio. It can be found at `https://github.com/carlosperate/awesome-microbit`.

Index

Packt.com

Subscribe to our online digital library for full access to over 7,000 books and videos, as well as industry leading tools to help you plan your personal development and advance your career. For more information, please visit our website.

Why subscribe?

- Spend less time learning and more time coding with practical eBooks and Videos from over 4,000 industry professionals

- Improve your learning with Skill Plans built especially for you

- Get a free eBook or video every month

- Fully searchable for easy access to vital information

- Copy and paste, print, and bookmark content

Did you know that Packt offers eBook versions of every book published, with PDF and ePub files available? You can upgrade to the eBook version at packt.com and as a print book customer, you are entitled to a discount on the eBook copy. Get in touch with us at customercare@packtpub.com for more details.

At www.packt.com, you can also read a collection of free technical articles, sign up for a range of free newsletters, and receive exclusive discounts and offers on Packt books and eBooks.

Other Books You May Enjoy

If you enjoyed this book, you may be interested in these other books by Packt:

Building IoT Visualizations using Grafana

Rodrigo Juan Hernández

ISBN: 9781803236124

- Install and configure Grafana in different types of environments
- Enable communication between your IoT devices using different protocols
- Build data sources by ingesting data from IoT devices
- Gather data from Grafana using different types of data sources
- Build actionable insights using plugins and analytics
- Deliver notifications across several communication channels
- Integrate Grafana with other platforms

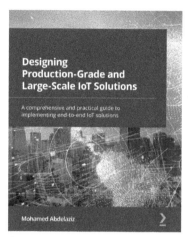

Designing Production-Grade and Large-Scale IoT Solutions

Mohamed Abdelaziz

ISBN: 9781838829254

- Understand the detailed anatomy of IoT solutions and explore their building blocks
- Explore IoT connectivity options and protocols used in designing IoT solutions
- Understand the value of IoT platforms in building IoT solutions
- Explore real-time operating systems used in microcontrollers
- Automate device administration tasks with IoT device management
- Master different architecture paradigms and decisions in IoT solutions
- Build and gain insights from IoT analytics solutions
- Get an overview of IoT solution operational excellence pillars

Packt is searching for authors like you

If you're interested in becoming an author for Packt, please visit `authors.packtpub.com` and apply today. We have worked with thousands of developers and tech professionals, just like you, to help them share their insight with the global tech community. You can make a general application, apply for a specific hot topic that we are recruiting an author for, or submit your own idea.

Share Your Thoughts

Now you've finished *BBC Micro:bit in Practice*, we'd love to hear your thoughts! Scan the QR code below to go straight to the Amazon review page for this book and share your feedback or leave a review on the site that you purchased it from.

`https://packt.link/r/1804610127`

Your review is important to us and the tech community and will help us make sure we're delivering excellent quality content.

Download a free PDF copy of this book

Thanks for purchasing this book!

Do you like to read on the go but are unable to carry your print books everywhere? Is your eBook purchase not compatible with the device of your choice?

Don't worry, now with every Packt book you get a DRM-free PDF version of that book at no cost.

Read anywhere, any place, on any device. Search, copy, and paste code from your favorite technical books directly into your application.

The perks don't stop there, you can get exclusive access to discounts, newsletters, and great free content in your inbox daily

Follow these simple steps to get the benefits:

1. Scan the QR code or visit the link below

https://packt.link/free-ebook/9781804610121

2. Submit your proof of purchase
3. That's it! We'll send your free PDF and other benefits to your email directly

www.ingramcontent.com/pod-product-compliance
Lightning Source LLC
Chambersburg PA
CBHW060520060326
40690CB00017B/3331